DICTIONARY OF
PLANT PATHOLOGY

DICTIONARY OF
PLANT PATHOLOGY

by
Dr. R.P. Chandola

2025
Daya Publishing House®
A Division of
Astral International Pvt. Ltd.
New Delhi – 110 002

First Published, 2010
Reprinted, 2025

ISBN: 978-81-7035-651-6

Publisher's Note:

Every possible effort has been made to ensure that the information contained in this book is accurate at the time of going to press, and the publisher and author cannot accept responsibility for any errors or omissions, however caused. No responsibility for loss or damage occasioned to any person acting, or refraining from action, as a result of the material in this publication can be accepted by the editor, the publisher or the author. The Publisher is not associated with any product or vendor mentioned in the book. The contents of this work are intended to further general scientific research, understanding and discussion only. Readers should consult with a specialist where appropriate.

Every effort has been made to trace the owners of copyright material used in this book, if any. The author and the publisher will be grateful for any omission brought to their notice for acknowledgement in the future editions of the book.

Published by : **Daya Publishing House®**
A Division of
Astral International Pvt. Ltd.
– ISO 9001:2015 Certified Company –
4736/23, Ansari Road, Darya Ganj
New Delhi-110 002
Ph. 011-43549197, 23278134
e-mail: info@astralint.com
Website: www.astralint.com

Printed at : **Replika Press Pvt. Ltd**

Preface

Words in any language have always taken a major role in the advancement of human civilization. With the passing of time many words automatically added and enriched the various cultures towards provisions of a better, secured and longer life of all the living beings on this planet–the earth. Words have been used in the area of education which in turn has played a significant role towards improvement and upgradation of human life. In this regard scientific knowledge acquired through modern scientific research and education has influenced the human life tremendously vis-à-vis all other connected human activities.

Agricultural science including animal science as a particular way of life as adopted by human beings and also moved in the same direction as other sciences and also acts made good progress only primarily by means up many words with passing out of time. As far as scientific development for further improvement is concerned, it firstly evolved the scheme of making dictionaries to make it possible for easy comprehension of human behaviour, actions and reactions towards a civilized society. Such works also helped and emerged as one of the most important way in achieving an easy consultation for an achievement of rapid and significant knowledge of the different disciplines of agricultural science. In the present vastness of ocean of knowledge of any subject, it is difficult and cumbersome find the correct words for the different functions of agricultural science. The situation has become more tedious as the worlds pertain to the scientific and technical notations of various agricultural branches of knowledge have become numerous.

With the allure in view, six volumes relevant to the various subjects of agriculture *e.g.* Animal Science, Entomology (insect pest study), Crop Science (growing various agricultural field crops, agronomy), Plant Science (plant breeding and genetics), Plant Pathology (plant diseases), and Soil Science have been prepared with all scientific names (terms) used in agricultural education and research with comprehensive notes for various terms. Whenever possible some entries have been illustrated to make an easy understanding possible. Such a dictionary should also be used as a quick and easy reference to students and other concerned groups.

Dr. R.P. Chandola

A

Abnormalities A plant develops and functions in a normal manner, if, however, it diverges from the normal and the abnormalities are sufficiently extensive, as to impair its very existence temporarily or permanently, then it may be considered as a disease shown by indications of certain abnormalities.

Acervuli The imperfect forms in Ascomycetes reach the culmination of their development by forming gemmae, oidia and several kinds of conidia. Conidia are borne on conidiophores, which may be scattered or more often closely packed together in a subepidermal or sub-cortical layers, abutting on one another in a continuous manner. These layers are called acervuli, if they

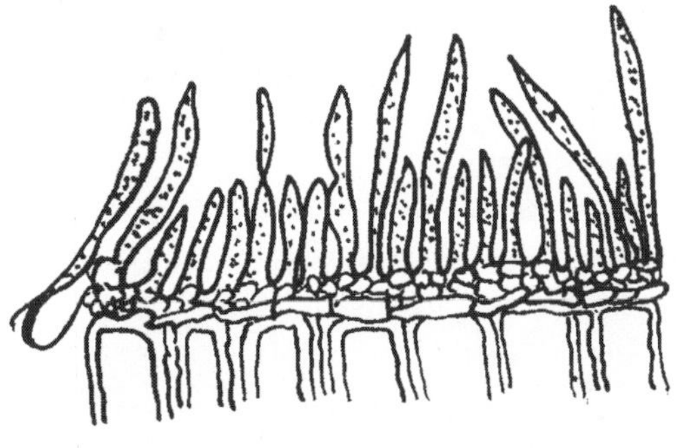

Acervulus of Colletotrichum Showing Setae

are saucer-shaped. Sporodochia, if are barrel-shaped or sausage-shaped. Conidiophores are enclosed in flask-shaped structures, the pycnidia, in which case the conidia are known as pycnidiospores. Pycnidia are provided with opening at their apical end known as ostioles.

Acinobaccillosis This disease is wooden tongue of cattle. This is caused by the organism *Actinobacillus lignieresei.* The disease attacks the throat-tissue of the cattle but is also often seen in the tongue, stomach, lungs and lymph glands. *See* Animal Science.

Acrogenous Denoting as during the formation of conidia, the basal conidia is the oldest.

Actinobacilliosis *See* Animal Science.

Actinomycosis This fungal disease causes Lumpy jaw in cattle. It is confined to the bones of the lower jaw, although the lower jaw and the nasal bones may also be involved. The causative organism for this is *Actinomyces bovis. See* Animal Science.

Aeciodoid teliospores Teliospores are produced as in the Uridinales in the genera Endophyllum and Kunkelia. They resemble the aeciospores were formerly mistaken for them. On germination, however, they produce four celled pro-mycelium. Such teliospores are known as aecidio-teliospores or micro-teliospores.

Aeciospores In Uridinales (fungi), reproductive structures after the mycelium has fully developed. The Uredinales are as a rule, polymorphic and different names have been applied to different stages, depending on their biological relationships and other malformation and their behaviour. The pycniospores (pycnidiospores); aeciospores (aecidiospores), urediospores (uredospores), teliospores (teleutospores) and sporidia (basidiospores). Except the sporidia, all other spore forms are borne in definite sori.

Aecium In *Puccinia,* the aecium is usually bell-shaped and is enclosed by one cell, thick wall known as peridium in cutting aeciospores, the uppermost cell with its daughter nuclei becomes the aeciospore mother cell.

Aerotropism Fungi respond to physical and chemical stimulti. They grow in response to moisture (hydrotropism), oxygen (aerotropism) or chemicals (chemotropism). Such response may be positive or negative, and different at different periods of growth. Broadly speaking, there are three periods of growth in the life of a fungus–(1) the vegetative phase (2) phase of a sexual reproduction and (3) the phase when sex organs are formed. These phases are inter-related but conditions optimum for one phase are not the same as for another.

Aetiology Science to find cause of disease.

Agallol A fungicide used to control charcoal rust of potato.

Agrosan GN An important seed dresser chemical–chlorophenol-mercury-phosphate.

Air borne *See* smut's infection.

Air-bone diseases For protection against diseases caused by air-bone fungi whose spores alight on plant leaves or stems, the application of chemicals to the foliage is the most effective way of disease control. The chemical prevents the germination of spores or kills the germ tube before it penetrates the host tissue if the spore has already germinated. Though many fungal and bacteruak diseases can be controlled in this way, chemicals cannot control virus diseases. In their case, protective measures have to be indirectly aimed at controlling the insect vector or by use of disease-free material for propagation, etc.

Albuginaceae It is a family of widespread distribution with a single genus, Cystopus, the species of which are obligate parasites on higher plants. The mycelium in unseptate, freely branched, and provided with globular, rarely knob-shaped haustoria.

Albugo The genus is called *Cystopus* also. It derives its name due to the formation of mealy white patches on the surface of the affected host. It is an obligate parasite fungus (endoparasitic), on most conifers and has also been reported on the members of Amaranthaceae, Compositeae, Convolvulaceae and Crucifereae. This has many biological races or formae speciales. Each race is restricted to an individual host species.

Albugo candida *See* white rust of mustard.

Allomyces Anisogamous heterogamy has been observed in this genus. Here the anthredia and oogonia are borne in short catenulate series on the same branch, one alternating with other, and distinguishable from each other by the relative density of their protoplasts.

Alternaria blight of linseed (*Linum usitatissimum*) Caused by *Alternaria lini.* Dark brown irregular patches on leaves which curl and dry up; diseased buds fail to open and are covered by dark brown spots which ultimately result in shrinkage of floral structures. Treat seed with Agrosan GN before sowing; grow late varieties.

Alternaria blight of mustard (*Brassica* spp.) Caused by the fungus *Alternaria brassiceae.* Brown circular spots on all parts of plant; spots in concentric rings with sooty black colour. Treat seed with hot water 50°C for 10 minutes; practice long rotation. Spray seed beds with 1 per cent Bordeaux mixture at the pod formation stage.

Alternaria brassiceae *See Alternaria blight* of mustard.

Alternaria brassicola *See* leaf spot and blight of *Brassica* spp.

Alternaria disease of Safflower (*Carthamus tinctorium*) Caused by *Alternaria carthami.* Spots composed of light and dark brown concentric rings on leaves; spots coalesce and form large, irregular lesions resulting in brittle texture; less attack on stems and petioles.

Alternaria lini *See* Alternaria blight of linseed.

Alternaria palandui *See* blight of garlic.

Alternaria palanduili* and *A. Porri *See* blight of onion.

Alternaria solani It is also known as early blight of potato. The disease first appears on leaves. They are scattered irregularly and in certain conditions they remain small and more or less angular being limited to the veins. Climate and soil exert a controlling influence upon the development of the disease. It becomes serious when the season begins with abundant moisture followed by high temperatures unfavourable to the host plant. *See* also leaf spot of Chillies and potato.

Alternaria tennuis Leaf spot of sunflower.

Alternaria triticina *See* leaf blight of wheat.

Amerosporae The form–genera of fungi imperfecti. They have spores one celled, round or egg-shaped, elongated.

Anguina tritici *See* air-cockle of wheat.

Angular leaf spot of cotton *See* Black arm disease of cotton.

Antheridia Morphologically differentiated male gametangia.

Antherozaid The male gamete of a fungus.

Anthracnose A plant disease caused by fungus *Colletotrichum gloeospordes.* It is attacking leaves, stems, inflorescences, flower and fruit of Citrus.

Anthracnose of Black Pepper (*Piper nigrum*) Caused by *Colletotrichum necator.* Brown to black lesions appear on mature leaves; spikes fall off premature; berries shrink and dry. Spray with 0.5 per cent Bordeaux mixture.

Anthrancnose of arecanuts and beetlenuts Caused by *Colletotrichum catechu.* Good manuring and cultivation.

Anthracnose of banana Caused by *Gloesporium musarum.* Spots on fruits. Spray with 1 per cent Bordeaux mixture.

Anthracnose of mango *Manigifana indica. Glomerella cingulata.*

Anthracnose and ripe rot of *Capsicum annuum* Caused by *Glomerella gossypi* and *Colletotrichum indicum.* All above ground parts attacked; infection on seedlings and bolls generally serious; and large red spots appear on seedling, which girdles the stem and kill it; spots are somewhat sunken on bolls; those on leaves produce scald effect; a slimy pink mass generally found in centre of spots. Treat seed with arganomercurials (Agrosan GN, Cerasan) before sowing @ 2–2.5 gms/kg; spray crop with 1 per cent Bordeaux mixture; destroy infected plants.

Anthracnose of cotton (*Gossypium* spp.) Caused by *Glomerella gossypi* and *Colletotrichum indicum.* All above ground parts attacked; infection on seedlings and bolls generally serious; and large red spots appear on seedling, which girdles their stems and kill them; spots are somewhat sunken on bolls; those on leaves produce scald effect; a slimy pink mass generally found in centre of spots. Treat seed with argano-mercurials (Agrosan GN, Cerasan) before sowing @ 2–2.5 gms/kg; spray crop with 1 per cent Bordeaux mixture; destroy infected plants.

Anthracnose of cowpea (*Vigna sinensis*) Caused by *Glomerella lindemuthianum.* Same as for *Phaseolus* spp.

Anthracnose of cucurbits Called *Colletotrichum lagenarium.*

Anthracnose of grape (*Vitis vinifera*) Depressed cankers form on stem and branches; brown spots with dark margin occur on leaves and brown sunken spots on berries. Paint stem immediately after pruning with ferrous sulphate solution in sulphuric acid (5 lbs. ferrous sulphate, half pint sulphuric acid and one gallon water); spray new shoots with 1 per cent Bordeaux mixture at two to three weeks intervals.

Anthracnose of horsegram (*Dolichos biflorus*) and *Phaseolus* spp. Caused by *Glomerella lindemuthianum.* Infected portions show dark brown sunken spots with raised reddish or yellowish margins; tissues of necrotic spots on leaves collapse become thin and papery; produces abundant pinkish slimy mass in humid weather; infected stems crack and rot.

Anthracnose of Lima beans (*Phaseolus lanatus*) Casual organism is *Colletotrichum lindemuthianum.* Elongated dark brown lesions appear on all portions above ground; lesions produce pinkish spores' masses during moist weather; veins, stems, petioles and pods bear spots; spots on hypocotyl cause death of plant; seedlings show canker on cotyledons; lesions on pods enlarge rapidly.

Anthracnose of *Phaseolus* spp. (*Phaseolus radiatus, P. mungo* etc.) Caused by *Glomerella lindemuthianum. See* Anthracnose of horsegram. Use healthy seed; practise crop rotation and sanitation, grow resistant varieties.

Anthracnose of sisal hemp (*Agave* spp.) Caused by *Colletotrichum agaves.* Innumerable black circular spots appear on leaves causing shrinking and drying of affected tissues; ultimately whole leaf dries up. Remove and burn affected leaves; spray crop with 1 per cent Bordeaux mixture.

Anthracnose of sword bean (*Canavalia ensiformis*) Caused by *Glomerella lindemuthianum.* Same as for *Phaseolus* spp. and *Colletotrichum capsici.*

Anthrax *See* Animal Science.

Anthridia When gamentangial copulation is between two morphologically differentiated gamentagia, then the male gamentangia are known as anthridia.

Aplanospores *See* Endogenous spores.

Apothecium It is a kind of ascocarp of saucer shape.

Apothecial initials Sclerotia of *Sclerotinia sclerotiorum* placed in moist sawdust at a low temperature in a dark room from only Apothecial-initials. They develop into apothecia as soon as they are exposed to bright light.

Apple blister A disease caused by fungus *Conothecium chomotosphoramin.* Can be controlled by removing the affected branches.

Applefore blight Caused due to a bacterium called Micrococcus anylomes.

Apple fruit rot A most common rot of fruit all over the world is the soft rot of blue rot caused by fungus *Penicillium expansum.*

Apple pink disease *See* pink disease. It occurs where climate is warm-moist and the trees are crowded. Rhinning of fruits reduces damage.

Apple-scab *Fusicladium dendriticum,* before it was known to be the imperfect stage of *Venturia inaequalis.*

Apple stem black Disease caused by *Conothecium cometosporium* fungus. Control depends on removal of diseased branch or by a paste containing 2 ozs. each of red lead and copper carbonate in 100 C.C. of raw linseed or spray twice in the spring season with lime-sulphur. The pruning would should be treated for stem-black with lanolin which is better than raw linseed oil.

Apple stem brown Disease caused by *Botryosphaeria ribis.* Ordinary pruning can control the disease.

Appressoria Ectoparasitic fungi have special outgrowth to serve as holdfasts or organs of attachments called appresoria.

Aquaculture Growing, marketing of the animals such as fishes, shrimps etc. for trade.

Aquatic fungi Many fungi grow well when immersed in water and some are normally found submerged in it. These are the aquatic fungi and belong as a rule to the Phycomycetes.

Arasan It is a non-metallic fungicidal dust chemically known as tetramethylthirum disulphide. It is used as a seed dresser.

Arch cells During the development of ascogonia, the distal ends are binucleate and the ultimate binucleate branchlet terminates is recurved to crozier-like (grooved) cell which is also binucleate. The two nuclei in this cell divide simultaneously and one pair lies in the arch of the crozier, known as the arch cell.

Aretan A fungicide used to control charcoal rust of potato.

Ascocarp It is the fruit body of higher ascomycetes. In fungi when the antheridium is functional or rudimentary, the ascogonium develops into a fruit in the body, called ascocarp.

Ascochyta Pathological usage has made the imperfect state, such as *Colletotrichum, Aschochyta, Cercospora,* the blight of gram is a various disease which very frequently appears in an epidemic form causing a loss upto even 50 per cent. If the weather during the period is dry, the disease remains restricted, but if there is rain at the time and the whole crops is involved, it may be completely destroyed in a few days.

Ascochyta leaf spot of Johnson grass (*Sorghum halepense*) Caused by *Ascochyta sorghi*. Same as for sorghum.

Ascochyta leaf spot of Sudan grass (*Sorghum sudanese*) Caused by *Ascochyta sorghii.* Same as of Sorghum.

Ascochyta pinodella *See* foot rot and blight of peas.

Ascochyta phaseolorum *See* Blight of beans.

Ascogenous cells The asci arise by the transformation of ascogenous cells, which are at first binucleate; but as the ascus matures, the two nuclei fuse, forming a large diploid nucleus.

Ascogenous hyphae The ascogonium instead of forming an ascus immediately, as in lower Ascomycetes, gives rise to one or many irregularly branched, tubular outgrowths known as ascogenous hyphae.

Ascogonia Same as oogonia. The female gametangia.

Ascomycetes This is one of the four classes of the Eumycetes or true fungi. In these, the sexual state is characterised by spores borne endogenously in asci.

Ascospore These are endogenously formed non-motile spores formed in an ascus. In Sphaeriales, the ascospores may be expelled through a pore at the apex of the ascus, or by rupturing of the ascus wall, or the upper half of the ascus may break. Ascospores lie in one (uniseriate), two (biseriate) or many (multiseriate) rows, or in a ball-like cluster in the ascus. In the ascus mother cell, the terminal cell and the penultimate cell ultimately fuse to form a binucleate cell just below the ascus mother cell. The two nuclei of the ascus mother cell fuse to form a diploid nucleus or synkaryon. The synkayron now divides thrice to form a haploid nuclei, the first division being meiotic. Each haploid nucleus gathers a small mass of cytoplasm and develops into a uninucleate ascospore.

Ascus The endogenous non-motile ascospores are formed in the Ascus.

Ascus mother cell *See* ascus formation when both the nuclei of the dykaryon in the hook. Both the nuclei of the dikaryon (in Aspergillus) in the hook cell undergo by the formation of a simultaneous division to form four nuclei followed by the formation of a uninucleate terminal cell, a binucleate subterminal cell or penultimate cell, and an antipenultimate cell forms the ascus proper, so it is called ascus mother cell.

Asexual spores These spores are accessory in character, designed for the rapid propagation and dissemination of a fungus. They are produced in indefinite numbers, always on the haplont mycelium, in the ascomycetes and, as a rule, on the diplont mycelium in the basidiomycetes.

Ashy stem blight of lima bean (Phaseolus lanatus) Caused by *macrophomina phaseoli.* Black sunken kanders appear just below cotyledonary nodes; lesions extent up the stem and kill growing point; disease advances more slowly, yellowing of leaves and stunting takes place and black sclerotia appear. Sow seed produced in disease free areas; application of wood ashers appears beneficial.

Aspergil-laceae Species belonging to this family are the common moulds found on all organic matter. They are as a rule saprophytes, but a few are animal parasites causing diseases known as mycoses. The importance of the members of this family is chiefly due to their use in industrial processes, especially the production of citric acid, gluconic acid, glycerol and drugs like penicillin and flavicin and in the ripening of cheese.

Aspergillosis *See* Animal Science.

Aspergilus One of the common bluish green mould (fungus). Most species live saprophytically on damp bread, jam, over ripe fruits, cheese, defectively preserved herbarium specimen and decaying organic matter. A few species are parasitic and become pathogens on animals and plants *e.g. aspergillus niger* caused to disease of dates, figs and others. *A. fumigatus* causes a disease called Aspergillosis in man and animals. Mycosis is also caused by this fungus.

The mycelium is hayline, bright or pale coloured, and bears on its surface concretions of colouring matter, but it is not by itself black or smoky brown.

Atthanni Fam. Rutaceae, *Citrus regulasa*.

Autocious In autoecious rusts, all the spore forms occur on the same host, *Melampsora lini,* which causes rust on linseed, is an autocious rust.

Autotrophic bacteria *See* Animal Science.

Avian ancepholitis Disease of young chicks. *See* Animal Science.

Avian chlamydiosis A disease of birds. *See* Animal Science.

Avian encephalomyelitis *See* Animal Science.

Avian influenza *See* Animal Science.

Avian leucosis complex *See* Animal Science.

Avian mycoplasmosis A bacterial disease of poultry. *See* Animal Science.

Avian pasteurelosis *See* Animal Science.

Avian tuberculosis Disease of chicken. *See* Animal Science.

B

Bacidium of rust

Bacillus botulinus Bacteria causing spoilage of fruits.

Bacteria *See* Animal Science.

Bacteria in relation to livestock raring *See* Animal Science.

Bacterial blight of bean (*Phaseolus* spp.) Caused by *Xanthomonas phaseoli indicus*—Large irregular, dark sunken areas, often brick red or brownish in colour, usually covered with yellowish white crusts of exudate; appear on pods; long reddish streaks on stem; plants look wilted. Use of disease free seed and practice of crop rotation are recommended for control of disease.

Bacterial blight of cowpea (Vigna sinensis) Casual organisms is the bacteria *Xanthomons vignicola.* Infected leaves show light yellow, irregular to circular spots with necrotic brown centre, later changing to straw column. Dark green water soaked spots of variable shape and size appear on pods which later become yellow and dry; affected leaves often fall off early, to practise crop rotation is the only control measure.

Bacterial blight of paddy (*Oryza sativa*) Caused by *Xanthomonas oryzae.* Water-soaked translucent streaks appear on leaves, which become yellow to white in colour, affected leaves dry and die.

Bacterial disease of alfalfa or lucern (*Medicago sativa*) Caused by bacteria *Xanthomonas alfalfae.* Small water soaked spots surrounded by a straw coloured halo appear on leaves, which turn brown later.

Bacterial disease of Sesbania Caused by *Xanthomonas sesbaniae.* Circular, water soaked pots surrounded by a halo appear on lower surface of leaf; upper surface chlorotic with brown centre; vertical streaks develop on stem and rachis from which pearly, gummy beads exude. Destruction of diseased plant material is the method by which this be controlled.

Bacterial disease of Soyabean (*Glycine max*) Caused by *Xanthomonas phaseolisojense.* Lesions common on leaves as yellow pustular outgrowths; later spots change to reddish-brown with marginal yellowing; small reddish brown spots also occur on pod.

Bacterial disease of sweet clover (*Melilotus* sp.) Cuased by *Xanthomonas alfalfa.* Same as for alfalfa.

Bacterial ear rot of wheat (*Triticum aestivum*) Caused by *Corynebacterinum tritici.* Leaves of affected plants curl; a bright yellow slimy ooze appears on leaves and inflorescence agglutinised inflorescence; and not seeds formed; grains converted into small hard galls. Control is same as for ear cockle.

Bacterial leaf spot of cabbage (*Brassica olearacea*) Caused by *Xanthomonas compestris var. armoreaceae.* Water-soaked leaf spot of minute size which with age may increase to 4–6 mm. Drying up of the leaves.

Bacterial leaf spot of chillies (*Capsicum annuum*) Caused by *Xanthomonas vesicularia.* Small, dark, greasy spots are formed on leaf, petiole and tender parts of stem. Water-soaked spots appear on green fruit. Centre of the spot becomes light brown and slightly sunken. Hot water treatment of seeds before sowing. Diseased plants should periodically be removed.

Bacterial leaf spot of cluster beans (*Cymopsis tetragonoloba*) A bacterial disease caused by *Xanthomonas cyamopsidis.* Small circular, water-soaked spots formed on leaves, which later become jet black, coalesce and form irregular lesions surrounded by chlorotic areas.

Bacterial leaf spot of mango (*Mangifera indica*) Caused by *Pseudomonas mangiferae-indica.* Minute water-soaked lesions bound by veins are formed on leaf blade. These lesions turn dark brown with age and are surrounded by chlorotic halo.

Bacterial leaf spot of tomato Caused by *Xanthomonas vesicatoria.* Same as for chillies.

Bacterial leaf spot of velvet bean (*Mucuna* sp.) Caused by *Xanthomonas stizolabiccola.* Lesions found on leaves as angular, livan sluscint dots which enlarge and become dark brown at a later stage. Crop sanitation is the control method.

Bacterial wilt of brinjal (*Solanum melongena*) Caused by *Pseudomonas solanacearum.* Lower leaves droop before wilting; vascular system becomes brown, same as for potato. For control other susceptible crops such as potato, tomato, chilli and tobacco should not be used in rotation; grow resistant varieties.

Bacterial wilt of tobacco (*Nicotiana* spp.) Caused by *Pseudomonas solanacearum.* Lower leaves droop and wilt; brown decay of pith; slimy bacterial matter exudes when stem is cut. Same as under potato.

Bacterial wilt of tomato Caused by *Pseudomonas solancerum.* Lower leaves droop; development of adventitious roots enhanced; wilt and death follow. Rotation with non-susceptible crops; sulphur lime treatment applying sulphur to soil to bring down the pH to 4 during summer followed by addition of lime to bring pH to 5.

Bacterial wilt, Ring of potato (*Solanum tuberosum*) Caused by *Pseudomonas solonacearum.* Affected plants develop bronze colour, shrivel and wilt; avoid replanting in infected soil. Use clean cultivation and disease free seed. Where cut potatoes are planted, knife should be dipped in a solution of mercury chloride (1–500).

Bactericidal substances Spraying plants with germicidal (bactericidal) substances is sometimes are effective in some crops.

Bactericidal substances in soil Considering the influence of soil organisms on higher plants is the capacity, enzymic or otherwise, that certain soil bacteria, fungi and actinomyces possess of inhibiting the growth of other micro-organisms, especially those not commonly found in soils. The failure of many disease organisms to survive in moist soils may, in part, be accounted for in his way.

Bacterium Non-motile rods.

Bacterium citri *See* citrus canker.

Bacteriology Study of micro-organisms like bacteria. Also called microbiology.

Baculovirus oryctes A microbial agent which successfully combats the menace of rhinocerous beetle of coconut palm.

Bajra green ear disease of Pennisetum typhoides Caused by the fungus *Sclerespora graminicola.* It is also called the downy mildew of bajra.

Bajra smut Caused by the fungus *Tolyposporium penicillariae,* only the ovaries are attacked, but not all of them in an ear are infected.

Bakanac disease of paddy *See* foot rot disease of paddy.

Balansia oryzae *See* Udbatta disease of paddy.

Banded leaf and sheath blight of maize Caused by *Thanatephora sacchari.*

Bangle disease of potato Same as bacterial wilt of potato.

Bannora bunchy top Caused by a virus.

Basidia *See* basidiomycetes.

Basidiomycetes These are divided into three classes depending on the manner in which basidia are formed. In one method, a terminal cell of the dikaryotic mycelium forms a spherical structure known as a chlamydospore. This is the probasidium and forms a basidium (also known as promycelium) on the germination of the former. This basidium may be septate or non-septate. Basiospores are formed on sterigmata on the basidium and they may produce secondary spores by budding. In the second method, the proba obasidium is known as a teliospore. This, on germination, develops a four celled basicium, each cell of which produces a single diospore. There is however no budding. In the third method, the basidium is formed directly on the terminal cell of the hypha. This is septate and several such basidia are arranged in a palisade-like layer in the mycelium.

Basidiospores *See* basidiomycetes.

Basidium The sex spores are formed exogenously on a basidium, and are called basidiospores.

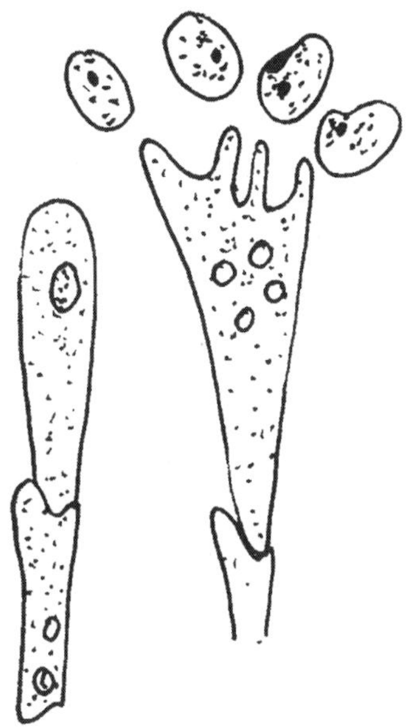

Basidium and Basidiospores

Bean rot (*Phaseolus* spp.) Caused by *Phytophthora parasitica.* Water-soaked lesions appear on pods, on which a white felt of mycelium develops and rotting takes place. Rogue infected pods; spray with 1 per cent Bordeaux mixture.

Bean rust Caused by *Uromyces appendiculata* Pers. On beans (Phaseolus vulgaris).

Belannhore indica *See* flowering parasite of coffee.

Berberis vulgaris In *Puccinia graminis,* pycnia and aecia are produced on this plant, the barberry, whereas the uredia and telia are produced on wheat plant.

Biological control of diseases These methods help in reducing the dependency on chemical fungicide. Species of *Trichoderm* and *Glioch* are being used as bioagents against several soil and seed borne diseases.

Biological race A strain which though morphologically indistinguishable yet differs from the normal in its pathogenicity.

Biological regulation *See* Crop Science.

Biotechnology Where biological methods of crop production are used as use of compost in the integrated pest management as to stop the use of chemical aspect of crop production.

Biotrophs *See* obligate parasites.

Bird's eye spot of tea (*Camellia sinensis*) Caused by *Cercosporella theae.* Disease usually attacks plants in nursery in monsoon and defoliates them; minute black or brown spots appear on young flush; leaves develop chocolate brown patches and get distorted and crumpled. Spray with 2 per cent Bordeaux mixture.

Bird's nest fungi They belong to the subclass *Eubasidiomycetes.* Are characterised by a true basidium. Other close members are the mushrooms, toadstools, braket fungi, puff-balls and stink horns.

Bitter rot of apple (*Pyrus malus*) Caused by *Glomerella cingulata.* Immature fruit starts rotting which continues during ripening and storage. Remove affected fruits and cankered branches; spray with Bordeaux mixture (4.5–50).

Black arm of angular leaf spot of cotton (*Gossypium* spp.) Caused by *Xanthomonas malvacearum.* All above ground parts of plants are infested; lesions on leaves are angular and water-soaked, which turn dark brown having reddish and purplish margin when old; those on stem are long and dark coloured. Control same as for *glomerella gosypi* or anthracnose.

Black band disease of jute (*Chorchorus* spp.) Caused by *Diploidia coechori.* Dense black discoloured band appears a round stem, two to three feet above ground level; plants wilt and die.

Black bean disease of coffee (*Coffea arabica*) Physiological disease. Blackening of beans in fruit varying from a pinhead spot to wholesale rotting. Spraying with Bordeaux mixture may be helpful.

Black finger of banana (*Musa* spp.) Caused by *Macrophoma musae.* Raised lesions raised on fingers. Remove affected fingers; spray with 1 per cent Bordeaux mixture.

Black heart disease of potatoes It is a non-parasitic disease. Oxygen starvation of tubers in storage in poorly ventilated bins, particularly of tubers at the bottom of deep piles kept for a long period; tissue of cut tuber turns pink on exposure, later turning dark brown or black. Provide proper storage conditions with adequate ventilation.

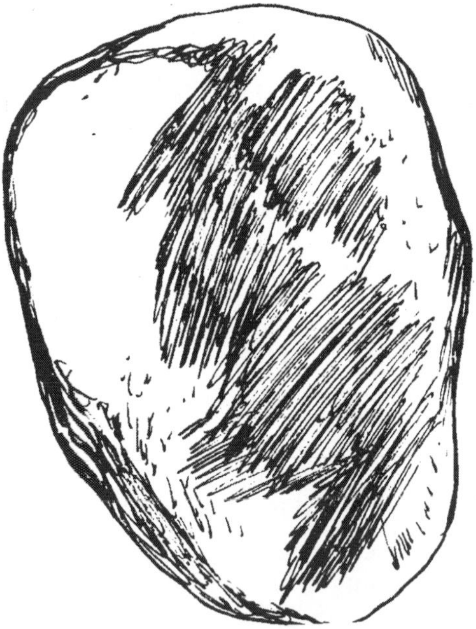

Black Heart Disease of Potato (Physiological)

Black leg of tobacco (*Nicotiana* spp.) Caused by *Phytophthora palmivora.* Black lesions are produced on stems starting from ground surface; leaves droop and fall over. Avoid over-irrigation; drench soil around plants with 1 per cent Bordeaux mixture.

Black mould of wheat (*Triticum aestivum*) Caused by the organism *Cladosporium herbarum;* greenish black velvety growth on glumes, stems and leaves.

Black quarter *See* Animal Science.

Black rot, canker and leaf spot of apple (*Pyrus malus*) Caused by *Sphaeropsis malorum.* Minute purple specks appear on leaves

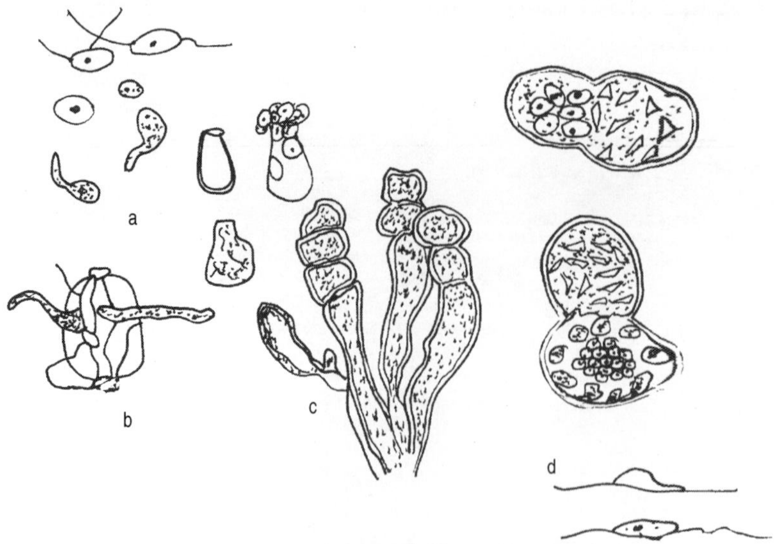

Stages in the Life of *Albugo candida*

a: Zoospores; b: Stomatal penetration by germ tubes c: Conidiophores; d: Oospore germination

which enlarge into spots; black rot appears a few weeks before maturity at calyx end; surface considerably wrinkled; branches and twigs produce canker or die back. Also causes fruit rot. Spray with lime sulphur (1 in 25) or Bordeaux mixture (4.5–50).

Black root rot of black pepper (*Pipper nigrum*) Caused by fungus *Rosellinia bunodes*. Roots are covered with a network of a black mycelium; white stars of mycelium found under the bark. Collect and burn all infected material; sprinkle Perenox solution (1½ oz. in 2 galls of water in hole before replanting.

Black root rot of jack fruit (*Artocarpus integrefolia*) Caused by *Rosillina* sp. Same as for tea.

Black root rot of cinchona Caused by *Rosellinia* spp. *See* for tea.

Black root rot of coffee Caused by *Rosellinia* sp. Same as for tea.

Black rot of coffee (*Coffee arabica*) Caused by *Pillicularia kolkeroga*. Under surface of leaves covered by thin pellicle of fungus extending through petioles; leaves, green twigs and fruits blacken and rot. Spray with Bordeaux mixture as mentioned in rust; spray again in breaks occurring during early part of monsoons.

Black rot of crucifers (*Brassica* spp.) Caused by *Xanthomonas (Pseudomonas) compestris*. Margin of leaves turns yellow; veins become dark; vascular discolouration of main stem; cabbage and cauliflower heads discoloured. Rotation with non-cruciferous crops for three years; treatment of seed with hot water at 50°C for 30 minutes.

Black rot of grape (*Vitis vinifera*) Caused by *Guignardia bidwelli*. Circular spots develop on half, mature berries, and minute black dots formed within; berries shrink and darken. Spray bunches several times with Bordeaux mixture or insoluble copper fungicides such as Cupraviti.

Black rot of mango Caused by fungus *Lasiodiploidia* spp.

Black rot of tea (*Camellia sinensis*) Caused by *Corticium invisum*. Tender leaves turn black, become soft and rot; local dead patches produced on older leaves in very wet weather; dead leaves remain attached to stem by a film of mycelium. Collect and burn prunings of affected bushes and those immediately around; spray with Bordeaux mixture after pruning.

Black root rot of tea (*Camellia sinensis*) Caused by *Rosellinia* spp. Affected roots are covered with a network of black strands of wooly mycelium; white starts of mycelium found under the root-bark. Collect and burn all infected material; dig a trench, 12 ft. deep, around the affected sport enclosing a ring of apparently healthy bushes; fork in lime over the affected patch; apply one gallon Perenox solution (1 oz. dissolved in two gallons of water) per hole before replanting.

Black rot of sweet potato (*Ipomea batatas*) Caused by *Ceralostomella fimbriata*. A storage rot of tubers, so occurs in field, yellow sickly appearance of foliage, and black cankers of portions underground; grey-black, circular, depressed spots on fleshy roots. Treat in the same way the black scurf of potato.

Black rust *Puccinia graminis*. It is very serious disease of wheat especially in southern and peninsular India. Also *See* stem rust of barley.

Black scurf of potato (*Solanum tuberosum*) Caused by *Pillicularia filamentosa.* Growing tips of tuber sprouts are killed, causing irregularity in potato stand; production of stem cankers resulting in formation of aerial tubers; tubers affected superficially by development of brown to black patches (sclerotia). Shallow painting with minimum covering of seed tubers; seed tuber treatment in hot formaldehyde (one part of 40 per cent formaldehyde in 120 parts of water at 50°C for 4 minutes), or acid mercury dip (1 part of mercuric chloride in 1000 parts of water and 1 per cent hydrochloric acid for 5 minutes; practice crop rotation.

Black tip Of mango–caused by the smoke of brick kilns. This hardens the tip of fruits which later turns black. Of banana–also called finger tip disease of banana fruits. It is caused by *Botrydiplodia* spp. It is most common in rainy season. Also is a disease of bananas.

Blacking and soft rot of potato (*Solanum tubersoum*) Caused by *Erwinia caratovora.* Affected plants turn pale green or yellow, wilt and die. Affected haulms are jet black in colour at soil level. Diseased tubers not in storage. Destroy diseased plants, wash potato with chlorinated water before storage.

Blast disease of paddy (*Oryza sativa*) Caused by *Piricularia oryzae.* Brown spindle shaped lesions on leaves; lesions on stem and grains darker in colour; rotting of neck and drooping of ears; affected grains partially filled. For control of the disease, treat seed with Cerasan or Agrosan GN before sowing @ 2–2.5 gms/kg; grow resistant varieties; burn affected straw and stubbles; spray with 1 per cent Bordeaux mixture.

Blast of finger millet (*Eleusine coracana*) Caused by *Piricularia oryzae.* Same as for paddy.

Blast of Italian millet (*Setaria italica*) Caused by *Piricularia oryzae.* Same as in paddy.

Blight (early) The important host in potato and is cuased by *Alternaria solani.*

Blight (fire) A disease of pears and apples caused by the bacteria *Micrococcus amylovorous.*

Blight (grey) It is caused by *Pestolotia theae* and is a harmful Blight disease of tea plant. As a rule, only old bushes are attacked.

Blight of bean (*Phaseolus* spp.) Caused by *Ascochyta phaseolorum.* Innumerable large brown spots with a discoloured margin of leaves; later, black minute dots appear on necrotic lesions arranged in circles. For control of the disease, use disease free seeds; destroy infected plant debris; spray with 1 per cent Bordeaux mixture.

Blight of brinjal (*Solanum melongena*) Caused by *Alternaria solani.* Leaf spots are dark and leathery; older leaves die prematurely and drop. Spray with 1 per cent Bordeaux mixture.

Blight of garlic (*Allium sativa*) Caused by *Alternaria palandui. See* under onion. Same as under Onion.

Blight (zea) Caused by the fungus *Gibberella zeae.*

Blight (late) A very harmful disease of potato caused by the organism *Phytophthora infestens.*

Blight of gram (*Cicer arietinum*) Caused by *Mycosphaerella rabiei.* Brown spots on stem, leaves and pods, these later become straw

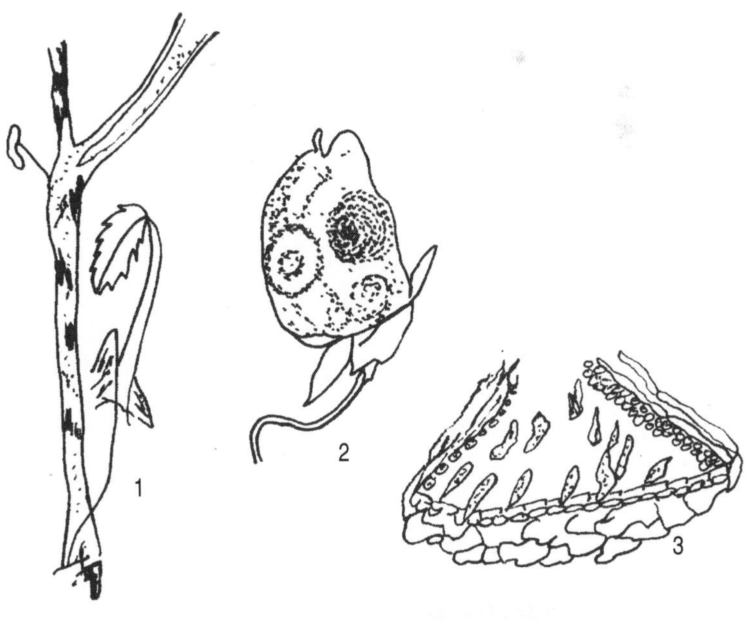

Blight of Gram *Ascochyta rabiei*

1: Affected plant showing lesions; 2: Affected pods; 3: Pycnidium showing conidiospores and conidia

coloured in the centre and studded dot-like, dark-brown fructifications, arranged in concentric circles. Control by practising crop rotation and sanitation, grow resistant vars.

Blight of gram ***Ascochyta rabiei*** (1) affected plant showing lesions, (2) affected pods, (3) pycnidium showing conidiophores and conidia.

Blight of onion (*Allium cepa*) Caused by *Alternaria palanduii* and *A. porri;* small sunken lesions with purple centre on leaf; lesions enlarge rapidly girdling leaf and flower stem which fall over; bulb decay follows in storage. Spray crop with 1 per cent B.M.

Blight of paddy (*Oryza sativa*) Caused by *Cochliobolus miyabeanus.* Innumerable dark brown elliptical spots on leaves, stem and glumes; inflorescence in mature plant becomes distorted; grain fail to form. Treat seed with Cerasan or Agrosan GN before sowing @ 2–2.5 gms/kg; spray crop with 1 per cent Bordeaux mixture; grow resistant varieties.

Blister blight of tea (*Camellia sinensis*) Caused by *Exobasidium vexans.* Small, yellow spots are formed on leaf; upper surface develops a shallow depression; lower surface becomes a greyish-white spore bearing convex; during wet weather, disease spreads to young green stems. Spray with Perenox (6 oz. to 15 gallons per acre), Cuprokylt or Blitox or dust with 5 per cent Cuprosana (10 lb. per acre) every week or ten days.

Blister disease Caused by the fungus *Conothecium chomatasporum.* It is a serious disease of apples and is also called stem-black disease of apples.

Blossom end rot of tomato (*Lycopersicum esculentum*) Caused by non parasitic organisms. Water-soaked spots at blossom-end of green or ripening fruit which enlarge quite rapidly, depressing the surface of lesion and become dark in colour. Avoid irregularity of moisture at time of luxuriant growth.

Borassus flabellifer Toddy-palm is badly infected by *Phytophthora palmivora. Cocus nucifera* (coconut palm) is also attacked by this blight.

Bordeaux mixture A popular fungicide. It is a mixture of copper sulphate and lime mixed in the proportion of ½ kg of each made to a gallon of water. Must be used freshly prepared to get good results.

Bordeaux mixture preparation Bordeaux mixture is prepared by mixing solutions of copper sulphate and quick lime in certain proportions. For preparing a 4–4–50 Bordeaux mixture, 4 lb. copper sulphate and 4 lb. quick lime are dissolved in 50 gallons of water. At first, stock solutions (1 lb. in one gallon) are prepared, and the two mixed together, just before use, by pouring copper sulphate solution into lime solution that has been diluted with remaining quantity of water. Copper sulphate may be dissolved by suspending it in water in a gunny sack in an earthern or wooden vessel. Lime is first slaked with a small quantity of water and made into a thin paste before the remaining water is added. The final reaction of the mixture should be either neutral or slightly alkaline, which can be tested with litmus paper or by dipping a clean iron nail in it which should remain free of any copper deposit. Bordeaux mixture deteriorates on keeping, hence only fresh preparations should be used. However, the deterioration can be avoided appreciably by addition ot cane sugar to the mixture at the rate of 60 grams for 50 gallons.

Botryodiploda theobromae *See* stem canker of Cinchona.

Botryodiplodia theobromae *See* root rot of tea.

Botryophaeria pruni-spinosae *See* die back of peach and apricot.

Botrytis infestans The name given to phytophthora infestans by Montague in 1845.

Botrytis viticola The downy mildew of grapes. It is now named as *Peronospora viticola*.

Botulism *See* Animal Science.

Bovine mastitis *See* Animal Science.

Branchy top of sesamum *See* Phyllody disease sesamum.

Brachy forms Some rusts may produce pycnia, uredia and telia but aecia are eliminated from their life cycles. These are branchy forms.

Branch canker and brachy forms Bright orange coloured pustules on both surfaces of leaves, surrounded by necrotic zones; later, reddish-brown to black pustules on stem resembling pustules of black rust. Grow resistant varieties; treat seed with Agrosan GN before sowing @ 2 gms/kg in the hilly tracts.

Branch canker of tea (*Camellia sinensis*) Caused by various fungi. Gnarled formations develop in young branches and on upper surface of old branches. Prune off cankered branches, scrape out dead wood and apply Bordeaux paste. Treat pruned cuts with coal tar, give an extra dose of potash in manure.

Braxy *See* Animal Science.

Broom rape of brinjal (*Solanum melongena*) Caused by *Orobanche cernua* and *O. indica*. One or more fleshy whitish shoots containing many fleshy scales and light blue flowers appear at base of affected plant; parasite is attached to root of host plant. Practise crop rotation; avoid susceptible crops such as tobacco and weeds like Datura; prevent setting of seed by pulling out rapes before they are mature and form seeds.

Brown blight of tea (*Camellia sinensis*) Caused by *Colletotrichum cameliae.* Yellowish green spots appear on young leaves and stems; gradually turning dark brown; when old, upper surface of leaves becomes a mixture of irregular patches of grey and red brown. Prune all dead and discoloured wood; spray immediately with Bordeaux mixture and again ten days later.

Brown eye spot of coffee (*Coffea arabica*) Caused by *Cercospora coffeicola.* Small circular spots on leaves causing leaf shedding. Reduce shade in nursery; spray lower surface of leaves every fortnight with ½ per cent Bordeaux mixture.

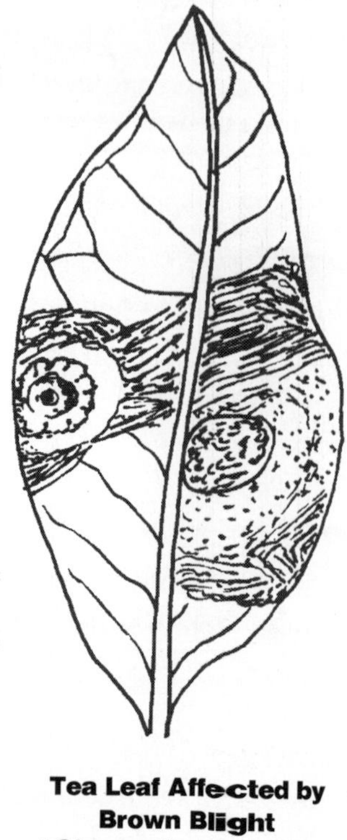

Tea Leaf Affected by Brown Blight (*Glomerella cingulata*)

Brown leaf spot of grape (*Vitis vinifera*) Caused by *Phakora viticola.* Dark brown angular spots on leaves and young shoots; regular manuring; spray with 1 per cent Bordeaux mixture.

Brown leaf spot of sugarcane (*Saccharum officinarum*) Caused by *Cercospora lingipes.* Young show narrow, oval spots on both surfaces; later, spots turn brown, red and yellow with concentric rings; still later they become broad and oval with straw coloured centre.

Brown root rot disease of tea *See* Formes noxious.

Brown or orange leaf rust It is wheat rust caused by the fungus *Puccinia triticina.*

Brown rape of tobacco (*Nicotiana* spp.) Caused by *Orobanche cermua. See* as under brinjal.

Brown root disease of cinchona Caused by *Formes noxious. See* in tea.

Brown root disease of jack fruit (*Artocarpus integrefolia*) Caused by *Formes noxious.* Same as for tea.

Brown root disease of tea (*Camellia sinensis*) Caused by *Formes noxious.* Leaves turn yellow and wither; roots are covered with white and brown mycelium which penetrates and forms a thin layer between bark and wood. In case of early stage of infection, expose main roots, scrap affected bark and drench with Aretan or Ceresan. Apply Prowex over the treated roots.

Brown rot A gummosis caused by *Phytophthora* spp. In Citrus.

Brown rot gummosis of Citrus fruits Caused by *Phytophthora palmivora* and *P. parasitica.* Profuse gumming of trunk bark near ground level girdling of tree; extension of infection to crown and roots. Avoid contact of bark with wet soil; apply Bordeaux mixture paint to trunk; dust holes at planting time with mixture of zinc sulphate, copper sulphate and lime (5 : 1 : 4); scrape affected bark and apply a wash of zinc-copper-lime mixture.

Brown rust of wheat Also called leaf rust, caused by *Puccinia reconditta.*

Brown spot of maize (*Zea mays*) Caused by *Physoderma zeamaydis.* Light green oval, slightly water-soaked lesions on leaves and leaf sheaths, later changing to reddish brown; stalks often break at infected points. For control, practise crop rotation and field sanitation.

Brown stripe downy mildew of maize Caused by *Sclerospothora rausiae.*

Brucellosis *See* Animal Science.

Bud-rot of toddy-palm Caused by *Phytophthora palmivora* and coconut palms.

Bud rot, root rot and nut fall of coconut (*Cocos nucifera*) Caused by *Phytophthora palmivora. See* Arecanut.

Bud sport A somatic mutation or variation arising in a bud and producing an abnormal branch. Also called bud son.

Buller phenomenon In *Puccinia helianthi,* the haplont hyphae of the mono-sporidial origin are known to anastomose with the uredial hyphae that are dikaryotic. The later thus contributes to the former the nucleus which is lacked and diploides it. Such diploidization of a haploint mycelium by a ikaryotic mycelium is known as the Buller's phenomenon.

Bunchy top A serious disease virus of bananas. Where it occurs, the plants should all be removed, burnt and fresh planting material secured from an area free from the disease green streaks in petioles and along leaf veins; shortening of crown and stunted growth. Select healthy suckers for planting; control aphid vector with insecticides.

Bunga disease of sugarcane (*Saccharum officinarum*) Caused by *Aeginetia indica.* It is a root parasite, 2 ft. high, with rudimentary scaly leaves and purplish black flowers borne on long red and white stalks. Remove infected stools with parasite and burn; plough up badly infected area.

Bunt of paddy (*Oryza sativa*) Causal organism is *Neovossia horrida.* Grains transformed into black sooty powder; usually three-four grains affected in an ear; symptoms appear in mature grains only; can be controlled by growing resistant varieties.

Bunt of wheat (*Triticum aestivum*) Caused by *Tilletia foetida* and *T. caries.* Affected plants ripen, ears assume dark green colour grains changed to a black sooty mass covered by membrane,

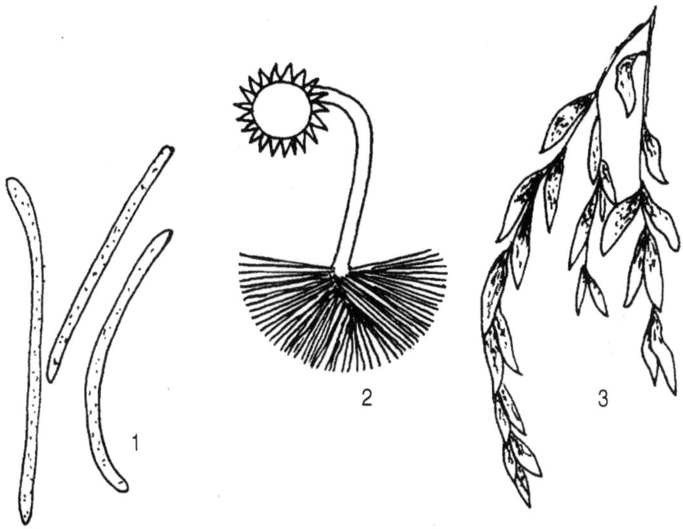

Bunt of Rice (*Neovossia horrida*)

1: Affected ear; 2: Germinating spores; 3: Sporidia

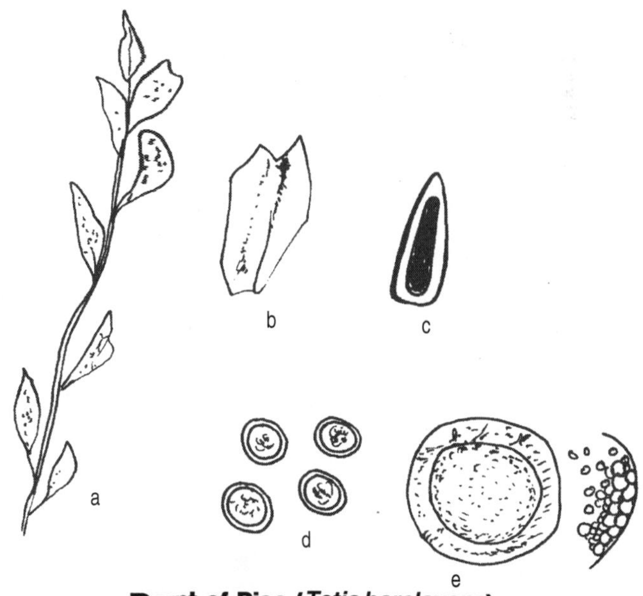

Bunt of Rice (*Tetia barclayana*)

a: Parts of bunted ears; b: Bunted grains;
c: Bunted grains longitudinal and cross section

affected grains smell of rotten fish. For control, treat seeds with Cerasan or Agrosan GN @ 2 gms/kg; grow resistant varieties.

Bunt of rice (*Tilletia barclayana*) (*a*) Parts of bunted ears, (*b*) bunted grains, and (*c*) bunted grains longitudinal and cross sections.

Burgundy mixure Burgundy mixture is a variation of Bordeaux mixture in which lime is replaced by sodium carbonate (washing soda). The method of preparation is just the same as for Bordeaux mixture. The usual strength of the solution is 4–6½–50. There are a number of proprietary copper fungicides available in the market *e.g.* perenox, prelan, blitix, Cuprokylt, Cuprosan, Fytolan, etc., which are as efficacious as and in some cases more so than Bordeaux mixture. The proprietary products also contain dispersing and wetting agents which make them more affective. A 2 per cent may be prepared with 5 kg copper sulphate, 6.5 kg washing soda made in 50 gallons of water.

C

Caceous lymphodinitis *See* Animal Science.

Calf dephtheris *See* Animal Science.

Campylobacter *See* Animal Science.

Campylobacteriosis Disease of man and chicken. *See* Animal Science.

Candiadisis *See* Animal Science.

Cankers Parasitic fungi attack the bark of plants, causing the death of tissues. When the bark sloughs off, open wounds are formed which are often of a spreading nature. They may be so sometimes surrounded by a raised, tumour-like margin and are known as Cankers. Cankers may encircle the entire plant, killing the upper part of the host. This happens in the pink disease of rubber caused by *Pellicularia salmonicolor,* or in apple trees stacked by *Physalospora mutila.* In some disease such as the stem bleeding disease of coconuts, there is no canker formation, but the bark may crack and a liquid oozes out.

Canker of citrus Caused by *Phytomonas citri.* One of the most prevalent disease of citrus stems, leaves, twigs and fruits. It is caused by a bacterium known as Bacterium citri. Use of lemons or Rangpur lime as stock is more resistant. Fruits and leaves may be kept covered with some fungicide such as Bordeaux mixture for prevention.

Canker and die back of mango (*Mangifera indica*) Leaves and twigs develop dieback. Prune diseased branches and spray with 1 per cent Bordeaux mixture.

Capsid The protein coat of a virus, surrounding the nucleic acid. A capsid is present in the inert extracelluar stage of the life cycle. Capsids are composed of sub-units called capsomeres. *See* also virus.

Cardamoms A serious virus disease causing leaf curl in this crop *(Elattaria cardamom)* causes heavy damage. Use of resistant varieties is the only remedy against this disease.

Castor blight of seedlings The causal organism for this disease on *Ficinus communis* (castor) is *Phytophthora parasitica.*

Catenulopeora As an exception, the teliaspores of Catenulopsora germinate without rest.

Cause of disease When a parasitic micro-organism enters into the causal complex of a disease, it is commonly considered to be the cause of the disease.

Cephalosporium sacchari *See* wilt of sugarcane.

Cerasan An important seed dresser chemical, ethyl mercury.

Ceratostamella paradoxa A reddish exudate is a common feature of this stem bleeding disease of coconuts and areca nuts.

Ceratostomella paradoxa *See* pine apple disease.

Ceratilium desmium *See* rust of cotton.

Ceratostomell fimbriata *See* black rot of sweet potato.

Cerestomella paradoxa *See* stem-bleeding of coconut.

Cerotelium fici *See* rust of figs.

Cercospora kopkei *See* yellow leaf spot of sugarcane.

Cercospora longipes *See* brown leaf spot of sugarcane.

Cercospora nicotianae Frog eye leaf spot of tobacco.

Cercospora musae *See* leaf spot of banana.

Cercospora beticol *See* leaf spot of beet.

Cercospora canavaliae *See* leaf spot of sword bean.

Cercospora canescens* and *C. cruenta *See* leaf spot of cow pea.

Cercospora carthami *See* leaf spot of safflower.

Cercospora coffeicola *See* brown eye spot of coffee.

Cercospora devisii *See* leaf spot of sweet clover.

Cercospora dolichi *See* leaf spot of lablab beans and horse-gram.

Cercospora indica *See* leaf spot of pigeon pea.

Cercospora nicotiana This causes the 'frog-eye' spot disease in tobacco leaves, does not occur outside the area outlined by the spot.

Cercospora oryzae *See* narrow brown leaf spot of paddy.

Cercospora ricinella *See* narrow brown leaf spot of paddy.

Cercospora sesami *See* seal spot of sesamum.

Cercospora sojina *See* leaf spot of soyabean.

Cercospora sorghili *See* leaf spot of maize.

Cercospora stizolobii *See* leaf spot of velvet'beans.

Cercospora traversiana *See* leaf spot of fenugreek.

Cercosporella theae *See* Bird's eye spot of tea.

Cereal root eelworm of wheat (*Triticum aestivum*) and barley (*Hordeum vulgare*) Caused by *Heterodera avnae.* Causes stunting of plants, unthriftiness of affected plants. Poor root formation, roots de-shaped. Long term rotation is needed to eradicate the worms from the soil.

Charcoal rot of finger millet (*Eleusine coracana*) and maize and sweet potato (*Ipomea batatas*) Caused by *Macrophomina phaseoli.* Root and lower part of stems affected, large lesions occur on stem whee small black sclerotia are formed at the later stage; shedding of leaves and rotting of stem takes place. Practice crop rotation.

Charcoal rot of jack fruit (*Artocarpus integrefolia*) Caused by *Ustulina zonata.* Same as for tea. Slow storage rot, tissue becomes chocolate brown, later dark red brown and ultimately black sclerotia formed. Raise vines form tubers which are not grown in an infected locality.

Charcoal rust of potato (*Solanum tuberosum*) Caused by *Macrophomina phaseoli.* Roots become brown and rot. On the stem portion, bark becomes ash coloured. Black sunken areas are formed around the stem end of tuber. Early harvesting; treat tubers from affected crop with agallol or aretan. Grow resistant varieties.

Charcoal rot of tea (*Camellia sinensis*) Caused by *Ustulina zonata.* Disease attacks tea roots by contact from infected roots of nearby shade trees; bushes die attached. Cut down and uproot shade trees; dig a trench round over-grown trees; and cut all side roots.

Chirke of cardamom (*Elettaria cardamomum*) Caused by Virus. Light and dark green patches are observed on leaves. Grow resistant varieties.

Chlamydospore In this method of basidia formation, a terminal cell of the dikaryotic mycelium forms a spherical structure known as chlamydospore. This is the probasidium and forms a basidium (also known as a promycelium) on the germination of the former. This basidium may be septate or non-septate. Basiodiospores are formed on sterigmata on the basidium and they produce secondary spores by budding. In this second method, the probasidium is known as Teliospore. This on germination, develops a four-celled basidium, each of which

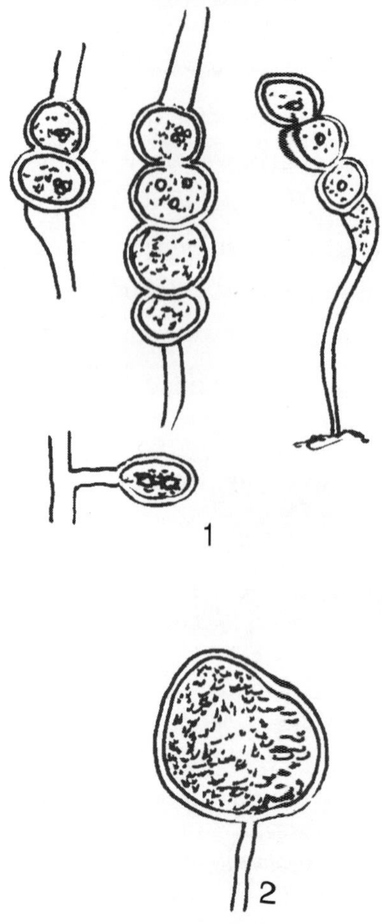

Chlamydospore of
1: Fusarium udum;
2: Phytophthora

produces a single basidiospore. There is however, no budding. In the third method, the basidium is formed directly on the terminal cell of the hypha. This is a septate and several such basidia are arranged in a palisade-like layer in the mycelium.

Choanephora cucurbitaceae A member of phycomycetes and is parasitic to sauces and chillies.

Choanephora indurasibulifera Causes rotting in flowering of *Hibiscus rosa-sinensis.*

Chyridiales This is an order of the sub-class Oomycetes.

Ciboria *See* Helotiales.

Cigar-end disease of banana (*Musa* spp.) Caused by *Stachylidium theobromae.* Distal end of fingers rot which, on drying give appearance of a lighted cigar. Remove pistil, stamens remains of perianth 8 to 11 days after bunch formation.

Citrus anthracnose On citrus. Withertip disease caused by *Colletrotrichum gloespoiodes,* the leaves.

Citrus canker Not all species of Citrus are susceptible to canker. *Citrus sinensis, Citrus grandis, Citrus poonensis* are usually not attacked. *Citrus jambhiri, Citrus aurantium* and *C. aurantifolia* are among the most susceptible species.

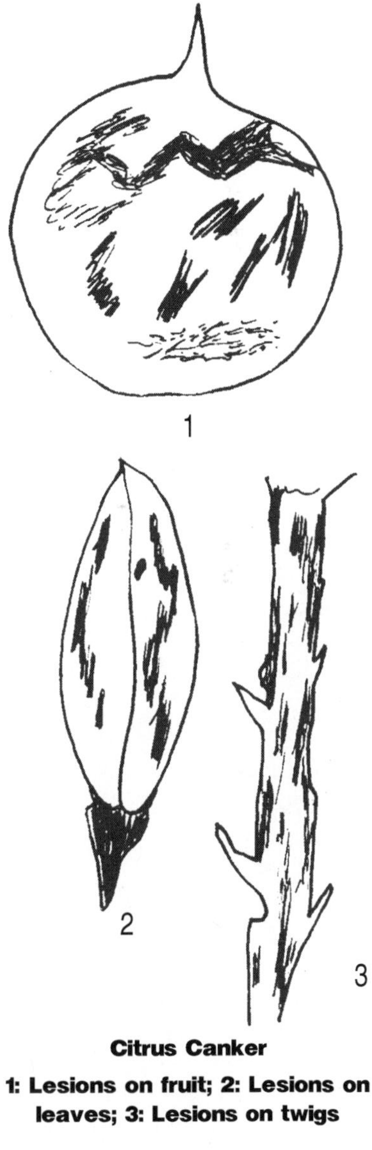

Citrus Canker

1: Lesions on fruit; 2: Lesions on leaves; 3: Lesions on twigs

Citrus die-back It is one of the most serious physiological diseases. In its advent, the leaves of a shoot turn yellow and fall, and shoot dies finally.

Citrus pink disease Of oranges. Caused by *Pellicularia (Corticium) salmonicolor*. It causes extensive damage to other plantation crops such as rubber, tea, coffee, cinchona, mango, etc.

Cleistothecium Bounded asci.

Cleistothecium *See* ascocarp.

Clavate *See* haustoria.

Claviceps purpurea Large sclerotia take the place of ovaries in Rye affected by this fungus.

Club rot, finger and toe disease of crucifers (*Brassica* spp.) Caused by *Plasmodiophora brassiceae*. Roots enlarge to form clubs (spindle shaped or of spheroid growth); temporary flagging of leaves on bright days which leads to wilt or retarding of growth. Avoid infested fields; lime the soil; treat seedling with mercuric chloride at transplanting time (1 in 1500) at the rate of 125 cc. (about 4 ozs.) per 100 seedlings.

Clump rot of cardamom (*Elettaria cardamomum*) Caused by *Pythium* spp. Rotting of rhizomes in clumps accompanied by death of leaves and falling over of pseudostem. Remove affected shoots; apply lime (3 oz. per clump).

Cluster bean rot (*Cyamopsis tetragonololia*) Caused by *Leveillule taurica var. macrospora*. Grey white fungus on lower surface of leaves. Same as for powdery mildew.

Coccidiosis *See* Animal Science.

Coffee rust (of leaf) It is caused by *Hemileia Vastatrix*. It so serious that at times it remains a major limiting factor in coffee production.

Coleosporium The name *Uredo oldendiae* was proposed first, but later on when its perfect state was found then it was transferred to Coleosporium and hence coleosporium old enlandiae. Its teliospores or an internal mycelium i.e. the teliospores become four celled and each cell is promycelial cell.

Colibacilloosis *See* Animal Science.

Collar rot Of apples caused by fungus *Rosellinia* spp.

Collar rot of groundnut (*Arachis hypogea*) White weft of mycelium appears at base of affected plant; numerous sclerotia, white to reddish tan, formed in mycelium, affected plants turn yellow and dry up. Grow resistant varies; soil may be disinfected with chloropicrin. Caused by *Phllicularia phaseoli.*

Colletotrichum Have their perfect state in physalospora for the perfect state of Colletotrichum. *Colletotrichum falcatum* is the casual organism of red rot of sugarcane. It is most serious and destructive. Colagenarium causes an anthracnose of cucurbits.

Colletotrichum agavae *See* Anthracnose of sisal hemp.

Colletotrichum camelliae *See* Brown light of Tea.

Colletotrichum capsici *See* leaf spot of Turmeric. Also *See* Anthracnose of Arecanuts.

Colletotrichum coffeanum *See* Die back disease of Coffee.

Colletotrichum gloesporiodes *See* wither tip of Citrus.

Colletotrichum graminicola *See* red leaf spot of sorghum.

Colletotrichum lindemuthianum *See* Anthracnose of lima bean.

Colletotrichum neccator *See* Anthracnose of Black pepper.

Colocasia bight It is caused by Phytophthora colocasiae and occurs only in rainy season.

Columella 1. The structure present in sporangia of many phycomycete fungi (*e.g.* mucor) produced by formation of a dome-shaped septum cutting off the sporangium from the sporangiophore. 2. The central column of sterile tissue in the sporangium of liverworts and mosses.

Common leaf spot of alfalfa (*Medicago sativa*) Cuased by fungus *Pseudopeziza medicaginea.* Small circular brown spot with black raised discs in centre appear on leaves; heavy infection causes defoliation.

Common yellow mosaic disease of Lima bean (*Phaseolus lanatus*) Caused by Virus. Causes various degrees of vein clearing and mottle on leaves; extreme malformation occurs in some varieties; general, yellow mosaic is more severe; varieties are susceptible. Grow resistant varieties; avoid close proximity of other host plants (such as sweet clover, red clover and gladiolus).

Conidiophore The stalk of conidia.

Conidiospore *See* conidium.

Conidium (Conidiospore) An asexual spore of certain fungi *e.g.* Pythium and Albugo. They are cut off extremely in chain at the apex of a specialized hypha, the conidiospore.

Coniothyrium arecae *See* leaf blight of Arecanuts and beetlenuts.

Coniugate nuclei Same as a diakaryons.

Control measures for plant diseases Cultural practices such as cooperation, mixed cropping, adjustment in date of sowing, depth of sowing and field sanitation through removal of diseased debris, weeds and alternate hosts play a great part in the control of plant diseases. A strict enforcement of quarantine, especially in the case of imported plants, helps to prevent the introduction of new diseases in the country.

As virus diseases cannot be prevented or cured by chemicals, use of disease-free seed goes a long way in reducing infection. For this, it is necessary to establish disease-free nurseries at isolated places where rigorous rouging eliminates all diseased material.

Copper blight of tea (*Camellia sinensis*) Caused by *Guignardia camelliae* and *Laestadia theae.*

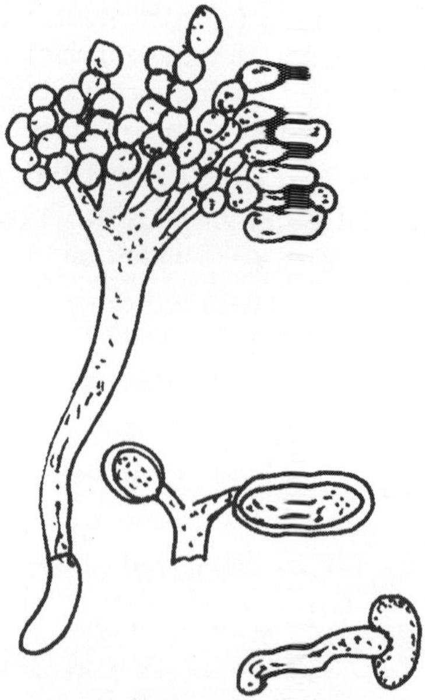

Conidiosphore and Conidia of *Sclerospora* sp.

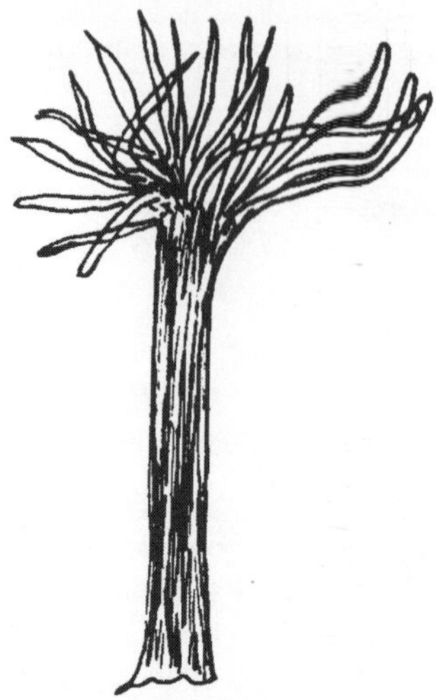

Conidiosphore of *Archrobotryum*

Large, irregular spots with minute black fructification points appearing on leaf; pick and burn affected leaves; spray with Bordeaux mixture and repeat after fortnight; after pruning, spray bushes with soda solution and first flush with Bordeaux mixture.

Copper fungicides The important copper fungicides are Bordeaux and Burgundy mixtures, and dispersible forms of cuprous oxide, copper oxichloride and basic carbonate.

Coremia When sporophores form clumps, they do so with or without lateral union. When these columns are in fascicles, they are called coremia.

Corticium invisum *See* Black rot of Tea.

Corticium rolfsii This is the perfect state of *Sclerotium folfsii* and occurs only when the dishes in which the fungus is growing are exposed to a higher temperature.

Corticium salmonicolor *See* Pink disease of jack fruit.

Coorynebacterium tritici *See* bacterial ear rot of wheat.

Cotton wilt Caused by *Fusarium vasinfectumi.* The disease is restricted to the black cotton soils.

Covered smut of oats (*Avena sativa*) Caused by *Ustilago kolleri.* Grains in ear of affected plants replaced by black mass of smut spores. For control of disease, treat seed with formalin or Agrosan GN before sowing @ 2.5 gms/kg seed.

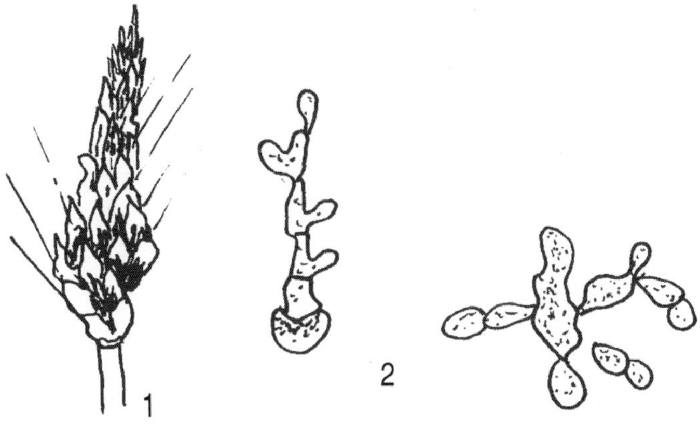

Covered Smut of Barley, *Ustilago hordei*

1: Affected ear; 2: Germinating spores

Covered smut of barley Caused by fungus *Ustilago hordi, Smutted heads;* franins replaced by agglutinised spore mass; covered by persistent white papery membrane. For control, treat seeds with cerasan or Agrosan GN before sowing @ 2–3 gms/kg of seed; grow resistant varieties.

Crinkling Disease caused by the virus. In these diseases, there are over-growths such as galls (as in Fijii disease of sugarcane), or lesions (sannhemp mosaic), stunting and dwarfing of plants, puckering and reduction in size of leaves and sterility are some of the other symptoms of this disease.

Crop protection Legislation has been enacted by the Government of India under Destructive Insects and Pests Act 1914, for the protection of crops, seeds, seedlings, etc. from destructive insect-pests and from other diseases both indigenous and imported. The Act also empowers Government to prevent the introduction and transport, from one state to another, of any insects, fungus or other pests destructive to crops. To enable the state governments to compel cultivations in particular regions where serious infestation has taken place, the Government of India have suggested that Pests Act should be amended with a view to doing away with the provision of appeal against Government authorities notifying areas where plant protection measures should be undertaken on a compulsory basis. In this connection, the U.P. Agricultural Disease and Pests' Act 1954 has been commended to the states.

Crown wart of alfalfa (*Medicago sativa*) Caused by *Physoderma alfalfae.* Dull coloured large galls formed on root at soil level, which turn grey or brown as they mature. Practise crop rotation, sanitation and proper drainage.

Crown wart of clover (*Trifolium* spp.) Casual organism is *Physoderma trifolii.* Same as alfalfa.

Crude rough Impure, unprocessed.

Cryptoccocosis *See* Animal Science.

Cultural races *See* differential hosts.

Cupravit An insoluble copper fungicide.

Curvularia penniseti *See* leaf spot of pearl millet.

***Cuscuta* sp. with its Host**

Cuscuta (dodders) Fam. Convolvulaceae. Thase are perennial, flowering parasites, 4 lbs granular CIPC; rain or irrigation needed for obtaining control of dodder in legumes especially in clovers, 4 lbs granular CIPC; rain or irrigation needed.

Cystidia Basidia form a more or less palisade-like hymenium and mingled with them in the hymenium Ire numerous, enlarged, sterile cells called the cystidia.

Cyst menatode of maize Caused by *Hetrodera zeae.*

Cystopus It is single genus in the Albuginaceae having a widespread distribution, the species are obligate parasites on higher plants.

In some species of Cystopus, more than one nuleus remain functional and are fertilized by the entry of an equal number female nuclei. The resulting oospores are thus multinucleate, and are known as compound oospores. The white rust caused by *Cystopus candidus* is the causal organism of the disease on various cruciferous plants like cabbage, turnip, mustard and radish.

Cystopus candidus *See* white rust of Brassica.

Cytospora sacchari *See* Stem canker of sugarcane.

D

Dacrymyces The basidiospores in the sub-class Eubasidiomycetes are as a rule unicellular, though in the Dacrymyces they may be transversely divided.

Damping off (1) Disease of many plants such as tobacco, where it does most of the damage. The most common organisms causing this disease are phytium de Baryanum and *P. aphanidermatum* and *P. myrotylum*. (2) The collapse of seedlings, ascribed to the attacks of such fungus organisms as *Botrytis vulgaris* and *Pythium* spp., in which plants wither, dry and die.

Damping off and foot rot of tomato (Lycopersicum) Caused by the fungi *Pythium* spp., *Phytophohora* spp. and *Ellicularia filamentosa.* Pre-emergence damping off of seedlings; post-emergence damping off relatively rapid; shrinking of cortical tissue of hypocotyl and falling over. Treat seed with hot water, with new Cerasan (2 per cent anhyl mercury phosphate) at 2 gms/kg of seeds against other organisms; store seed in loosely woven bags and not in air tight containers.

Damping off disease of brinjal (*solanum melongena*) Caused by *Pythium* spp., *Phytophthora* spp. and *Pellicularia filamentosa.* Causes rotting of seedlings; at ground level followed by falling over. Rotation of the seed bed; seed treatment with improved

Cerasan. For attack of *P. filamentosa*, hot water treatment of seeds (at 52°C for 30 minutes) followed by Cerasan treatment.

Damping off chillies (*capsicum annum*) Caused by *Pythium aphanidermatium*, *Phytophthora* spp. and *Pilicularia filamentosa*. Shrinking of cortical tissue of hypocotyle and falling over. Seed dressing with new improved Cerasan (0.05 per cent of weight of seeds). In case of Pillicularia, hot water treatment at 52°C for 32 minutes followed by dressing with Cerasan @ 2 gms/kg/ seed rotation of seed bed.

Damping off disease of cotton *See* wet rot of cotton.

Damping off in Lady's finger (*Abelmoschas esculentus*) Caused by *Pythium aphanidermatum* and *Phytophthora palmivora*. Seedlings attacked at ground level get constricted and fall off. Spray seedling with 1 per cent Bordeaux mixture.

Damping off of Pyrethrum (*chrysanthemum cinarariaefolium*) Caused by *Pythium* spp. post-emergence damping off in nursery. Thin out seedlings, lift excessive shade, and drench nursery with Bordeaux mixture.

Damping off of tobacco (*Nicotiana* spp.) Caused by *Pythium aphanidermatum* and *Phytophthora palmivora*. Constriction of seedlings at ground level which fall over. Apply 1 per cent Bordeaux mixture once a fortnight (500 gallons per acre of nursery) from the 20th day after sowing; crowded beds, use of raised sand-bed nurseries and rotation of nurseries are helpful.

Decline of citrus Caused by insufficient soil moisture and nutrition. Yellowing and gradual reduction of leaf; dieback of twigs followed decline and gradual death. Grow deep rooted green manure crops; avoid digging around trees; apply complete manure regularly every year.

Deep freeze Very low temperature.

Deficiencies disease It is known that plants suffer from deficiencies apart from nitrogen, phosphorus and potash. Among the deficient elements, though needed in very small quantities, copper, zinc, manganese, magnesium, iron and boron and molybdenum occur in an orchard.

Deodars' witches brooms These disfigurations in many cases are produced by parasitic fungi whose mycelium infests the internal tissue of the plant. These structures are witches' brooms and the casual organism for this is the *Peridermium cedri.*

Dermatitis *See* Animal Science.

Dermatophytoses *See* Animal Science.

Deuteromycetes The fungi imperfecti are called Deteromycetes because of their imperfectly known life cycle.

***Dianthus* spp.** These are attacked by Ustilago violaceae. They destroy the whole flower.

Dichlinous In phycomycetes, when anthredia are formed on the stalk of the oogonia or a branch arising from it, they are called androgynous. If they develop on independent branches, they are called dichlinous.

Dicotyosporae Among the fungi *imperfecti,* this family contains spores which are muriform, ovate and elongate, hayline or bright or dark or swarthy.

Dieback of apple Caused by *Botryosphaeria ribis.* Prune dead branches and give a bark-wash with Bordeaux mixture (4.5–50).

Dieback of coffee (*Coffee arabica*) Caused by *Colletotrichum coffeanum.* Leaves shed; twigs blacken and dry from tip downwards; berries dry up prematurely. Given an extra dose of manure when the heavy crop is maturing; provide good drainage.

Dieback of cowpea (*Vigna sinensis*) Caused by *Colletotrichum capsici.* Twigs and branches dry up from tip downwards, small, black, dot-like structures appear on dried up portions; infected pods shrivel. Sanitation and rotation of crops is required to control the disease. Also Bordeaux mixture (4 : 4 : 50) at 15 days interval, commencing as soon as disease appears.

Dieback of horsegram (*Dolichos biflorus*) Same as for cowpea.

Dieback of peach and apricoty Caused by *Botryosphaeria pruni-spinosae.* Young twigs develop dieback. Prune affected branches; with spray 1 per cent Bordeaux mixture.

Dieback of tea (*Camellia sinensis*) Caused by *Nectria cinnabarina.* Bushes become moribund and new shoots are thin and weak; small pink cushion-like fruit bodies appear in large numbers after the shoots die, followed by minute dark red spheres in small groups. Diseased bushes should be pruned back to good wood and sprayed with Bordeaux mixture.

Differential hosts Some races in the plant diseases can be grown on culture media and it is possible to distinguish races in them by difference in their growth, colour, consistency, topography, of the surface and other characteristics. Such races are termed as 'cultural races' to distinguish them from parasitic races, which are differentiated only by their capacity to infect certain species or varieties of plants and others.

Dikaryons *See* dikarotic phase. Related to fungi belonging to Ascomycetes and Basiodiomycetes

Dikaryotic-phase In the lower fungi, the conjugation of sex cells or the sex organs is normally followed by the fusion of the haploid nuclei, leading to the formation of zygote also called synkaryon. The process of fusion of the nuclei is known as karyogamy, the sex nuclei lie side by side in pairs and are termed conjugate nuclei or dikaryons. They initiate all vegetative activity synchronously, as though complete karyogamy had taken place. This phase, virtually diploid but containing two sexually differentiated haploid nuclei, is called the binucleate or dikaryotic phase.

Dimerosporae A family of fungi imperfecti identified by having spores two-celled, egg shaped or elongate, hayline or bright, dark or swarthy. Fungi may be monoecious, where both sexes occur on the same thallus and dioecious where the sexes occur on the different thali.

Diplanetism For a time, the zoospores are biflagellate and motile after which they become encysted by a wall. Later germinate by putting out germ-tubes and form a mycelium. In some species, zoospores have two periods of motility, a rest period intervening; this phenomenon is known as diplanetism.

Diplococus A form of bacteria.

Diploidia cajan A disease of *Cajanus cajan*.

Diplodia gummosis of *Citrus* spp. Caused by *Diplodia natalensis*. Gum oozes from forks and bigger limb; bark killed and gum dries on surface; sometimes infection spreads to wood. Scrape wounds and apply Bordeaux paint or zinc-copper lime paste.

Diploidia natalensis *See* diplodia gummosis of Citrus.

Diploidia corchori *See* black band disease of jute.

Diplophospora leaf spot of wheat (*Triticum aestivum*) Caused by *Diplophospora alopecuri.* Flecks on the leaves are formed which are yellow, elliptical-elongate, sometimes spindle like, later a black crust is developed in the centre. For control, cut tops and burn stubbles of affected plants. Treat seed with fungicides.

Disease avoidance Immunity or resistance may be due to avoidance or endurance of diseases or true resistance of disease.

Disease caused by bacteria (1) Human diseases such as cholera, diphtheria, dysentery, gonorrhoea, lockjaw, plague, pneumonia, syphilis, typhoid, tuberculosis and leprosy. (2) Animal diseases are Anthrax, black-leg of cattle, chicken cholera, pneumonia and tuberculosis. (3) Diseases of plants are fire-blight of apples and pears, Citrus canker, potato scab, ring-rot of potato, steward disease of corn, wilt of solanaceae and black rot of cabbage.

Dispersal of spores Fungi produce thousands of spores, so that some at least, if not at all, may find a suitable substratum and proper conditions of growth. To increase the number of spores and therefore the chances of survival of the fungus, several devices have developed among the fungi, chief among which are (1) production kinds of spores: polymorphism (2) increasing the spore bearing capacity of the same organ: chains of spores, multicellular spores, branching of sporophores and (3) increasing the spore-bearing surface, formation of gills.

Dissemination Devices for the dissemination of the spores are common in fungi, spores are liberated by the disintegration of the host tissue, by the hygroscopic movements of the sporophores, by ejection and other means that aid liberation. Their dissemination is affected by wind, water and insects.

Dormant mycelium This is the simplest resting mycelium of fungi.

Double streak *See* mosaic of tomato.

Downy mildew of alfalfa (*Medicago sativa*) Caused by *Peronospora aestivalis.* Greyish white powdered patches formed on under surface of leaves; corresponding portions of upper surface become yellowish and necrotic. Practise crop rotation and sanitation. Spray crop with 1 per cent Bordeaux mixture.

Downy mildew of clover (*Trifolium* spp.) Caused by fungus *Personspora trifolii repentis.* Same as for alfalfa.

Downy mildew of sweet clover (*Melilotus* spp.) Caused by *Peronospora aestivalis* and *P. meliloti.* Same as for alfalfa.

Downy mildew of common millet (*Panicum miliacium*) Caused by *Sclerospora gaminicola.* This fungus causes excessive tillering making the plants dwarf; further branches produced from auxiliary buds; floral reacts turn into leaf-like malformations; kernel fail to form, leaves show necrosis and browning and followed by splitting and shedding. For control, grow resistant varieties. Destroy the diseased plant debris.

Downy mildew of cucurbita Caused by fungus *Pseudoperonospora cubensis.* Purplish spots appear on lower surface of leaves; fruits do not mature or do not attain right flavour. Spray with 1 per cent Bordeaux mixture @ 0.6 per cent.

Downy mildew of fenugreek (*Trigenella foenumgraecum*) Caused by *Peronospora trigonellae.* Same as under *Phaseolus* spp.

Downy mildew of finger millet (*Eleusine coracana*) Caused by the fungus *Sclerophthora macrospara.* Spikelets of affected plants change into leafy shoots; remain upright and look like a brush.

Downy mildew of grapes (*Vitis vinifera*) It is a common disease of grape vines. This kills all the succulent parts of the plant, total destruction of leaves, dwarfing and killing of the shoots, and rotting and cracking of the berries. The causal organism are *Plasmapara viticola* and *Uncinula necato.* These produce powdery growth on both sides of leaf, young shoots, inflorescence and developing fruits. Dust with finely powdered (22 mesh) sulphur.

Downy mildew of Italian millet (*Setaria italica*) Same as for common millet.

Downy mildew of Johnson grass (*Sorghum helepense*) Caused by *Sclerospora sorghi.* Same as in Sorghum.

Downy mildew of Jowar Casual organism is *Sclerespora sorghi.* The disease is soil borne.

Downy mildew of lathyrus (*Lathyrus sativus*) Caused by *Peronospora lathyripalistris.* Leaves and twigs show downy growth, grey

Dry root rot of cluster beans (*Cyamopsis tetragonoloba*) Caused by fungus *Macrophomina phaseoli.* Same as in cotton.

Dry root rot of gram (*Cicer arietinum*) Caused by *Macrophomina phaseoli.* Same as under cotton.

Dry root rot of horse-gram Same as for Cotton and *Phaseolus* spp.

Dry root rot of lentil (*Lens esculent*) Caused by *Macrophomina phaseoli.* Same as for cotton.

Dry root rot for lady's finger (*Abllmoschos esculentus*) Caused by *Macrophomina phaseoli.* Outer parts of root shredded; tiny black sclerotia developed.

Dry root rot of cow pea (*Vigna sinensis*) Caused by *Macrophomina phaseoli.* Same as for cotton.

Dry root rot of *Phaseolus* spp. Same as for cotton.

Dry root-rot of sesamum (*Sesamum orientale*) Caused by *Macrophomina phaseoli.* Same as for cotton.

Dry root rot of soyabean (*Glycine max*) Caused by *Macrophomina phaseoli.* Small round or irregular black sclerotia appear below the epidermis on affected stem and root. For control, *See* as in cotton.

Dry root rot or sore thin of cotton (*Gossypium* spp.) This disease is caused by *Macrophomina phaseoli.* Large brown lesions formed at collar portion; vascular portions of roots discoloured brown; rootless rot and plant dies. Practise crop rotation, grow resistant varieties.

Duck hepatitis *See* Animai Science.

Duck septiaemia A disease of ducks. *See* Animal Science.

Duck virus hepatitis *See* Animal Science.

Dust fungicides By far the simplest methods of seed treatment is to use chemicals in dust form. The dust must be of extreme fineness, that is, capable of passing through a sieve with 300 meshes per linear inch; otherwise it will hardly be effective.

E

Ear-cockle of wheat (*Triticum aestivum*) Caused by *Anguina tritici*. Leaf blades are twisted, infection causes short size of ears prolong greenish-ness, awna are more wide spreading, affected grain changed to one or more small galls. For control, use of clean seed is advised. The seed should be free from galls. Galls should be removed by sieving or by floating the grain in water.

Early blight of brinjal (*Solanum melongena*) Caused by the fungus *Alternaria solani*. Oval or angular dark brown to black spots on leaflets forming concentric rings on dead tissue; lower leaves dry up and droop first; surface lesions on tubers are dark sunken and irregular in shape. Crop rotation for control; also spray crop with carbamates.

Early blight of tomato Caused by *Alternaria solani*. Circular to angular, dark brown to black spots on leaves which wither, droop and fall off; stems and fruits also affected. Disinfect seed with Zinebs and Naban.

Echinulation Non motole spores of fungi have variously sculptured outer membrane, the exospore or the epispore. Sculpturing may be in the form of warts, spines, echinulations, reticulations or rings.

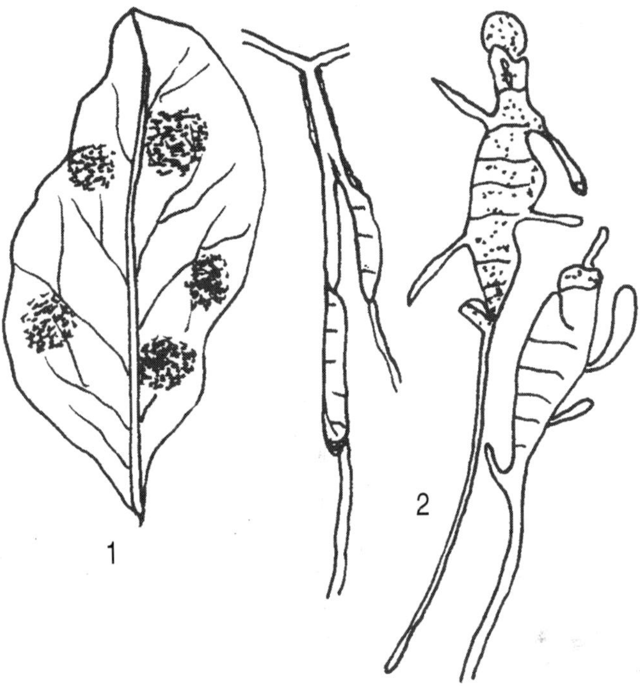

Early Blight of Potato: ***Alternaria solani***
1: Affected leaf; 2: Spores

Ectoparasite Parasite fungi mayor may not penetrate the host. If they grow on the surface of the leaves, stem and other parts; the mycelium is known as ectoparasitic.

Ectoparasitic When parasitic mycelium grows on the surface of leaves, stems and other parts. Ectoparasites have special outgrowth to serve as holdfasts or organs of attachment called appressoria.

Egg Female gamete.

Egg perinotis *See* Animal Science.

Elettaria cardemoni The cardomoms are affected by a serious virus disease.

Empusa A member of Phycomycetes of which the hyphae are characterised by their large diameter and fatty contents. They emerge from their insect hosts in white masses of characteristic appearance.

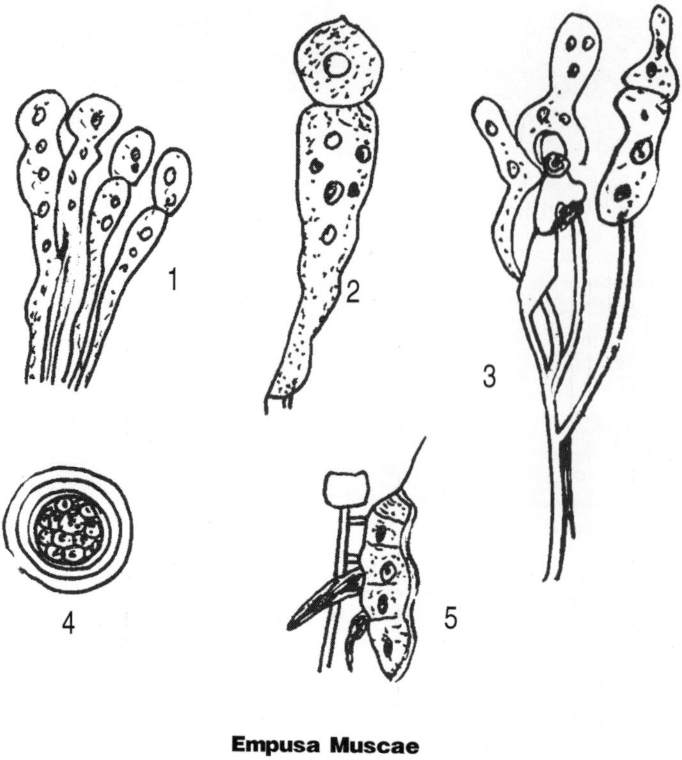

Empusa Muscae

(1) Group of condiophores showing conidia at various stages of development; (2) Conidiophore bearing conidium before discharge; (3) *Empusa grylli*, group of conidiophores of different ages; (4) A mature resting spore; (5) An acridian attacked by Empusa fungus.

Enation As seen in the sann-hemp mosaic.

Endogenous spores Spores are considered from the point of view of their morphology, biology and physiology. Morphologically the spores are endogenously or exogenously formed, motile or non-motile, one or more celled. Motile spores are known as zoospores or swarm spores. They are endogenously produced and the cell within they are fashioned is known as sporangium. Zoospores are naked and are provided with one/two flagella whose movements give them a circular or a swarming movement. Sporangia are formed at the end of the sprophores known as sporangiophores. Less often sporangia may be found intercalarily. In the course of evolution, the zoospores of some fungi have lost their flagella and thus their motility. Such endogenously produced, non-motile spores occur in mucor, and are known as sporangiospores or aplanospores.

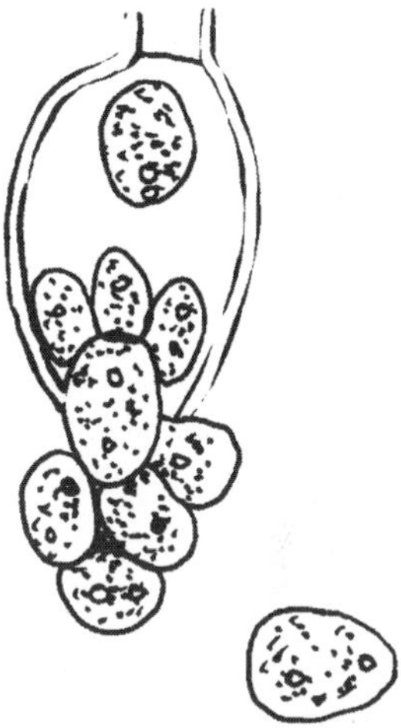

Endogenously Formed Spores, Zoospores of *Phytophthora parasitica.*

Endoparasite An organism that lives inside the body of another organism (the host) and benefits from the relationship at the expense of the host, such as the malarial parasite. Compare ectoparasite. *See* parasitism.

Endophyllum *See* Accidiod-teliospores.

Endophytic If the mycelium of a fungus is submerged into the substratum, it is endophytic.

Endoplasm *See* plasmasol.

Endospore Non-motile spores have a firm cell wall which is simple or manifold. In the latter case, there is a thin hylinex inner membrane–the exospore or the epispore.

Endosporic The formation of spores within the spore producing body; for example, the production of ascospores inside the ascus in ascomycete fungi. Compare exosporic.

Entamophthorales A subclass of Phycomycetes, of which three genera *viz.* Rhizopus, Mucor or Empusa are very important plant pathogens.

Entyloma oryzae *See* leaf smut of paddy.

Environment Factors of environment exercise a profound influence on the life of fungi, on their growth, reproduction, just as they do with other living organisms.

Enzyme These are secretion products of fungal and bacterial cell.

Epidemiology Study of factors affecting the outbreak of an infectious disease.

Epiphytic When the mycelium lives on the surface of the substratum.

Epiphytotic When a disease occurs in a severe/form epidemic in the ascus, a certain amount of unused cytoplasm known as endplasm.

Epizotic lymphangitis A disease of horse. *See* Animal Science.

Equatorial rainforests *See* Plant Science.

Eradication At one time, it was supposed that the eradication of the alternate host would help in eliminating black rust, because of the incapacity of the sporidia to infect wheat, and of the uredospores, which do infect wheat, to survive the summer temperature of the Indian plains. It has now been discovered that the urediospores serve as an inoculum in a more general manner than was formerly realised.

Eremascus fertilis In the lower ascomycetes, such as *E. fertilis*, sexual branches arise from neighbouring cells of the same hypha and grow side by side.

Ergot *See* sclerotium.

Erysiphales Are characterised by their obligate parasitism and by being typically ectoparasitic.

Erysiphe cichoracearum *See* powdery mildew of cucurbits. Also *See* powdery mildew of lady's finger.

Erysiphe polygoni *See* powdery mildew of peas.

Erysiphe Basigenous spore formation is the characteristic of this group. Conidia vary in shape, size and colour. If they are multicellular, they vary in the number of arrangements of cells. Sub-class hemiascomycetes, asci formed singly, directly from the zygote, sometimes closely aggregated, but no ascocarp formed.

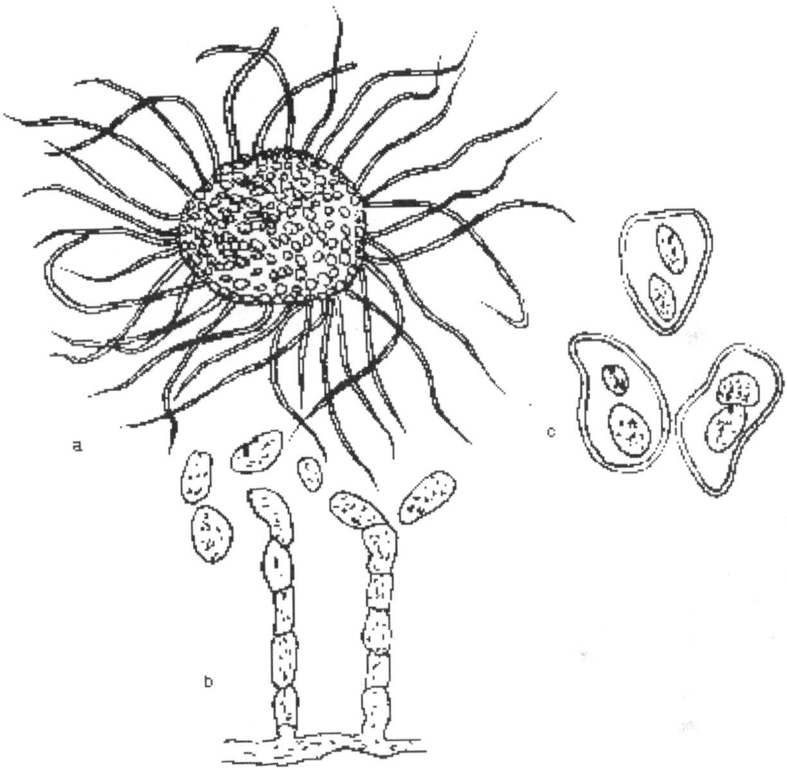

Erysiphe cichoracearum

a: Cleistheecium; b: Conidiophores; c: Asci and ascospores

Erysipelas A poultry disease. *See* Animal Science.

Erysiphe graminis *See* powdery mildew of wheat and barley.

Erysiphe cichoracearum *See* powdery mildew of tobacco.

Erysiphe polygoni Powdery mildew of coriander, cumin and poppy.

Euascomycetes A subclass of Ascomycetes in which asci are formed directly from a zygote from which more than one eascus are formed and asci borne in ascospores.

Eubacteria A large and diverse group of bacteria, principally distinguishable by their rigid cell walls that give the cells a fixed form. Most are unicells that are divided by binary fission. The cells can be spherical, rod-shaped, or helical. Most are immotile, but some possess flagella. They are a ubicutious group, some being found in extreme conditions. *See* bacteria.

Eubasidiomycetes A subclass of Basidiomycetes in which basidia are strictly limited growth, aseptate, bearing a definite number of basidiospores, usually four.

Eumycetes A class of Thallophyta and are the true fungi.

Eumycophyta Higher fungi.

Eutetix tenellus Curly top of sugar-beet is transmitted by the leaf hopper, but not by any other sucking insects feeding on sugar beet.

Exobasidium vexans *See* blister blight of tea.

Exogenous spores These are borne on various types of sporophores. The sporophore may consist of a single or branched hypha, or it may be bundle of hyphae. It has a vertical position, limited growth and a characteristic mode of branching.

Exospore A thick, often variously sculptured outer membrane.

Exosporic The formation of spores outside the spore producing organ. In the Basidiomycete fungi, for example, the basidiospores are born on the tips of the sterigmata, which are outgrowths of the basidium spore production, is more commonly endosporic, compare endosporic.

Exudation The production of a clear, amber-coloured exudation on the surface of the affected plant, which later sets into a solid mass insoluble in water, is termed gummosis. This disease is common in cherry, peach and citrus trees. Stem-bleeding disease of coconuts is due to *Cerestomella paradoxa*.

Eye spot of sugarcane (*Saccharum officinarum*) Caused by *Helminthosporium sacchari*. Elongated pale yellow leaf spots appear on both sides of young leaves, later turning red with yellow margin. Grow resistant varieties.

F

Facultative anaerobes A big number of bacteria are facultative anaerobes *i.e.* they can grow best in the presence of oxygen and can also live as anaerobes. In both the types, respiration heat is produced which is sometimes so enormous that stacks of grass or hay may catch fire. At times, this heat energy is converted into light energy and emits the phosphorescent glow. This phenomenon is technically known as bioluminescense and is responsible for the 'Ghost Light' of forest.

Facultative parasites This term implies that such organisms are ordinarily saprophytic but under certain conditions they may become parasitic.

Facultative saprophytes There are various categories of parasitism. These may subsist as saprophytes for apart of their life cycle away from the host. Many of the smut fungi are good examples of this.

Fairy rings The mycelium of the Basidiomycetes, living in the soil sometimes forms, in the Eubasidiomycetes, fairy-rings, sclerotia or rhizomorphs.

False smut of paddy (*Oryza sativa*) Caused by *Ustilaginoidea virens.* Affected grains transformed into large compact greenish-black mass; only a few grains in an ear affected. May be controlled by seed treatments.

Fames noxius *See* brown root disease of Cinchona.

Favus in chicken *See* Animal Science.

Fermate A new fungicide made from non-metallic synthetic compounds have been favourably reported, among which the fermate and the diathane appear to have given best results. Chemically it is ferric dimethyldiothiocarbamate.

Fig rust disease of fig leaves Caused by fungus *Uredo fici.*

Fiji disease of sugarcane This virus disease forms overgrowths such as galls.

Finger and toe disease of crucifers *See* Club rot of Crucifers.

Finger tip disease *See* black tip.

Fission bacteria The principal method of multiplication of bacteria is by simple fission.

Flag smut of wheat (*Triticum aestivum*) Caused by *Urocystis tritici.* Streaks of grey to grey-black colour on leaves; leaf sheaths and sometimes on stalks and glumes; when upper thin layer ruptures, it gives out black spore powder; plants stunted and leaves twisted; ear does not form. As control, grow resistant varieties; sowing shallow and good irrigation; crop rotation should be practised.

lavicin *See* aspergillaceae.

Flexuous hyphae The pycnium sends into such nectar numerous paraphyses and a small number of slender and branched hyphae.

Flowering parasite of coffee (*Coffee arabica*) Caused by *Balanophora indica.* Tuberous outgrowths parasitise lateral roots of coffee; floor heads put out in October–November; put deep isolation trench around affected area, and uproot and burn all coffee within the trench.

Foliar blight of wheat Caused by *Helminthosporium sativum* and *H. triticina repentis.*

Foliocellosis *See* mottle leaf.

Fornes noious *See* Brown root rot disease of Tea.

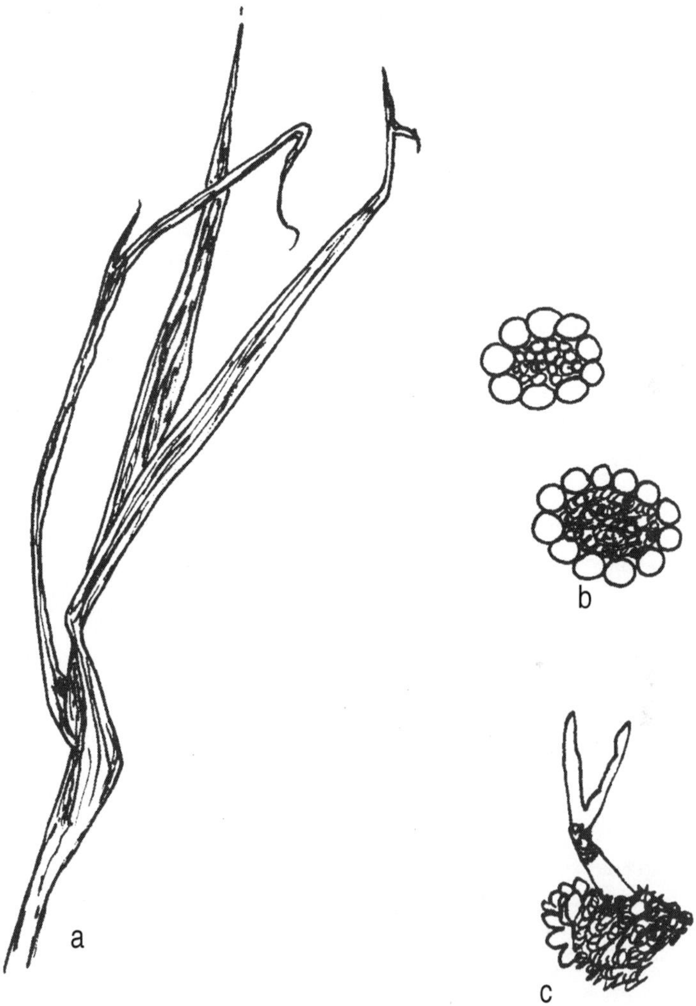

Flag Smut of Wheat

a: Part of affected shoot; b: Spore balls; c: Germinating spore ball

Foot The sporogonium of funaria at maturity is about 2–4 cms long and is divisible into three main regions–foot, Seta and capsule. Foot is short, poorly developed, acuminate type and is embedded in the stem tip from which it draws nourishment for the growing sporangium. It is not visible superficially and can be seen only in section of the stem tip.

Foot rot The stems at the level of the soil rots, though the parts above and below are more or less intact. This disease is caused by *Phytophthora* spp.

Foot rot and blight of peas (*Pisum sativum*) Casual organism are *Ascochyta pinodella, A. pisi, A. pinodes.* Small purple spots appear on leaves, which on enlarging turn brown to black; stem lesions are similar, but elongate and coalesce producing a blue black coat on lower stem; lesions on pods appear as tan, circular, sunken spots with dark margin. Grow crop in areas of low rainfall; practise crop rotation with no pea growing three to four years in the affected field; grow resistant varieties.

Foot rot disease of paddy (*Oryza sativa*) Caused by *Gibberella fujikuroi.* Affected plants stunted, late and fail to tiller; root become reddish brown and rot; grains shrivelled; some of the diseased plants appear taller than others. Controlled by treating seeds with Agrosan GN or Cerasan before sowing @ 2–2.5 gms/ kg, grow resistant varieties.

Foot rot of wheat Caused by *Sclerotium rolfsi.*

Foot rot and wilt of tomato Caused by *Pillicularia (sclerotium) rolfsi.* Same as blight of brinjal.

Foreign varieties Varieties of wheat resistant to loose/smuts have been obtained from foreign countries.

Form genera There are six families in the Monliales, four in the Sphaeropsidales and one in the Melanconiales. These families are not based on phylogenetic relationship but on superficial and artificial resemblance. The genera comprising them are therefore known as form-genera.

Fowl cholera An infectious disease of fowl. *See* Animal Science

Fowl coryza A disease of chicken. *See* Animal Science.

Fowl favus *See* Animal Science.

Fowl pox This virus is member of the genus *Avipox* of the family Poxvitidae. The incubation period is 4 to 10 days. The disease sings are mild with focal cutaneous lesions in recovered birds immunitare said to be life long. Different types of live vaccines are available.

Fowl typhoid *See* Animal Science.

Frenching *See* Mottle leaf.

Frong eye leaf spot tobacco (*Nicotiana* spp.) Caused by *Cercospora nicotianae.* Oval brown leaf spots occur in nursery and later in field. Treat seed with silver nitrate solution (1 in 1,000); spray plants in early stages with colloidial copper.

Fruit rot of apples By washing the fruits with 5 per cent solution of bleaching powder or 3 per cent of Mach-3. Remove the blotches before storage.

Fruit rot of bread fruit (*Artocarpus incisa*) Caused by *Phytophthora palmivora.* Same as in Jack fruit.

Fruit rot of Jack fruit (*Artocarpus integrefolia*) Caused by *Phytophthora palmivora.* Fruits develop soft rot. Spray with Bordeaux mixture.

Fruit rot, nutfall and bud rot of arecanut and betlenut (*Areca catechu*) Caused by *Phytophthora palmivora* (Koleroga disease).Tender nuts rot and fall off, but also rot. Spray bunches before monsoon (late May); with 1 per cent Bordeaux mixture; repeat during break of rains in July-August directing spray towards spindle of palm to protect buds.

Fruit rot of tomato (*Lycopersicum esculentum*) Caused by the fungus *Phytophthora* spp. Damping off and fruit rot; root appears as brownish spot, often at point of contact between fruit and soil, decay proceeds rapidly. Avoid poorly drained area; support plants with stakes; transplant on ridges in irrigated soils; spray with 1 per cent Bordeaux mixture.

Fungi A diverse group of mainly terrestrial organisms separated from other plants by their lack of chlorophyll. They are generally saprophytic or parasitic. The fungi are unicellular or composed of filaments (termed hyphae) that together comprise the fungal body or mycelium. Hyphae may grow loosely or form a compacted mass of pseudoparenchyma giving well defined structures, as in toadstools. The term fungi has no systematic significance, the true fungi being placed in the division Eumycophyta while the slime fungi form the division Myxomycophyta. *See* also Ascomycetes, Basidiomycetes, Phycomycetes.

Fungicides Such chemicals used to control fungus diseases of plants or animals.

Fungi imperfecti The perfect state of these fungi is unknown. A class of mostly microscopic fungi *e.g.* moulds, in which the sexual stage has not been discovered or where it has been lost in the course of evolution. Reproduction is by conidiospores. *See* deuteromycetes also.

Fungi in soil These are of three categories–(1) yeasts (2) moulds and (3) mushrooms. Of these, the last two are important in soils as yeasts occur to a very limited extent in such a habitat.

Fungi reaction to stimuli Effect of acidity and alkalinity (hydrogen-hydroxylation concentration) on the growth of fungi has shown that they tolerate a wide variations of hydrogen and hydroxylation fungus sclerotinia americana, apothecia develop in soils only with an acid reaction. Their development is inhibited if pH of the soils is 6.8 or more.

Fusarium annuum *See* wilt of chillies.

Fusarium arthoceros* var. *lathyri *See* wilt of Lathyrus root.

Fusarium congluinans Causes cabbage yellows. This is controlled by a single pair of gene because the resistance against this disease is governed by a single factor pair.

Fusarium coeruleum *See* dry rot of potato.

Fusarium cucurbitae *See* Fusarium root rot of cucurbits.

Fusarium indica *See* wilt of pigeon pea.

Fusarium lini More than two genetic factors appear to be concerned in deciding whether a variety is resistant or susceptible, to wilt due to *Fusarium lini*. Same is true for *Fusarium vasinfectum*, disease of cotton.

Fusarium oxysporium *See* wilt of tomato and banana.

Fusarium orthoceros *See* Fusarium gram wilt.

Fusarium root rot of cucurbits Caused by *Fusarium solani* and *F. cucurbitae*. Causes sudden mid season wilt; base of stem becomes dark brown, with a soft, mushy cortical wilt; fruits rot; young seedlings damp off. Hot water treatment of seeds at 55°C for 15 minutes, followed by treatment with mercuric chloride (1 in 1,000); avoid cucurbits for three years.

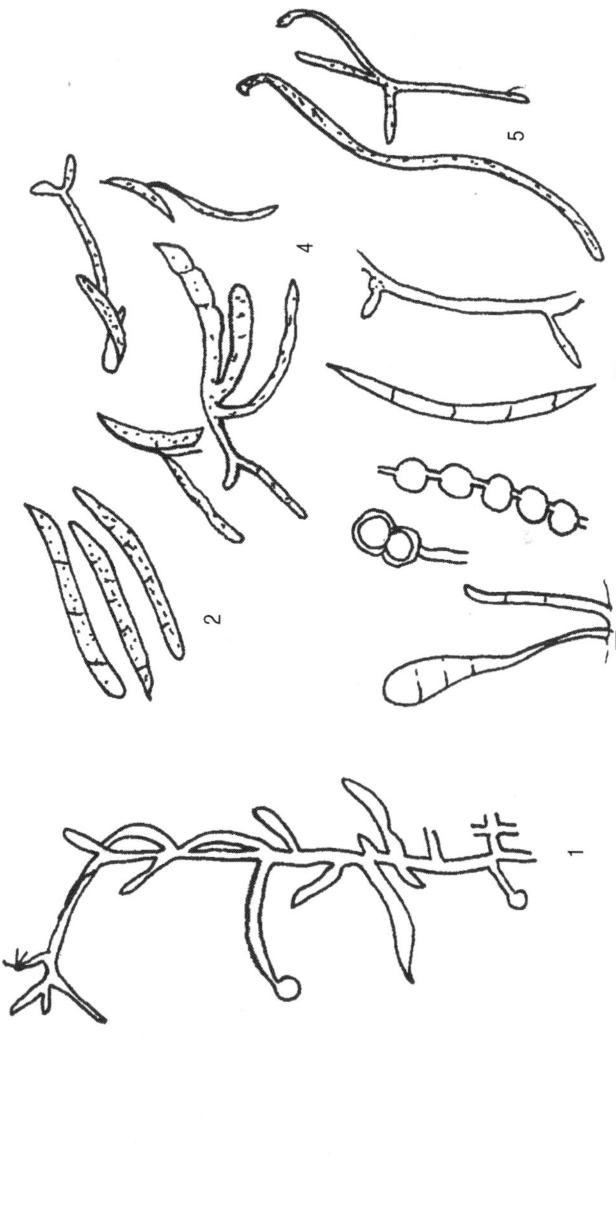

Fusarium udum

1: Mycelium showing microconidia; 2: Macroconidia; 3: Chlamydospores and their formation; 4: Germination of macroconidia; 5: Microconidia

Fusarium udum It is facultative parasite and can remain in the soil in a viable state for long periods of time. This is wilt of pigeon pea. In san-hemp wilt var. crotalkariae, the temperature for the optimum growth of the fungus is 25 to 28°C and coincides with the maximum development of the diseases.

Fusarium udum* var. *cyotolarae *See* wilt of san-hemp.

Fusarium vasinfectum This disease of cotton (*Gosypium* spp.) can be detected in the vascular tissues of the roots, stems, branches and in advanced cases in the petioles of leaves. *See* wilt of cotton. Also *See* Fusarium lini.

Fusarium wilt of gram (*Ciceur arietinum*) Caused by *Fusarium orthoeeros* var. *ciceri.* Affected plants stunted, with pale yellow green leaves, curled downwards; later, plant withers and dies; rootlets completely rotten; main root discoloured. For control, sowing late, deep ploughing and use of resistant varieties.

Fusarium wilt of tomato (*Lycopersicm esculentum*) Caused by *Fusarium oxysporium.* Yellowing of leaves which wilt and die and cling to stem; one or more branches also affected; plant remains stunted. Grow resistant varieties such as Marglobe, Pan American and Southland.

G

Gametangial copulation Sometimes in fungi, the formation of gametes is suppressed or eliminated and the sex act takes place between two gametangia themselves. This is more common sex act in fungi.

Gametangium The mother cell bearing the sex cells.

Gametes Sex cells arising within a mother cell.

Ganoderma lucidum *See* root rot of Arecanuts. Also *See* root rot of coconut.

Gangrenous dermatitis A bacterial disease of chicken-broilers. *See* Animal Science.

Ganoderma One of the orders of Eubasiodiomycetes.

Gemmae In Ascomycetes, the imperfect forms reach the culmination of their development in this class. Gemmae, oidia and several kinds of conidia are known, and in some species several imperfect forms may appear successively or simultaneously.

Gibberella fujikuroi *See* foot rot disease of paddy.

Gibberella zeae It is responsible for blight of maize.

Gills Spores may not be formed on all parts of the sporophore. In mushrooms, they are developed on specialised structures the gills, on the underside of the cap.

Ginger rhizome rot The rhizomes are completely rotted by *Pythium myriotylum* and *P. aphanidermatum*.

Gleba A closed structure in which the portion is enclosed is also found. In such sporophore, the surrounding membrane is known as a peridium and the basidia, forming portion as gleba.

Gloesporium ampelophagum *See* anthracnose of grape.

Gloesporium musarum *See* anthracnose of banana.

Glomerella cingulata Causal organism of the leaf spot of tea. Also *See* bitter rot of apples.

Glomerella lindemuthianum *See* Anthracnose of cotton, cowpea, horsegram and *Phaseolus* spp.

***Glomerella tucumanesis* (*Collectotrichum falcatum*)** *See* red rot of sugarcane.

Clomerulus A bunch of capillaries enclosed in the Bowman's capsule.

Goldern nematode of potato *Heterodera rostochiensis*. Stunting and unthriftiness of affected plants. Application of nematocides to disease infested plots. Long rotation should be used.

Grain smut of Johnson grass Same as for Sorghum.

Grain smut of jowar Grain smut caused by *Sphacelotheca sorghi* is the most common and destructive disease of sorghum.

Grain smut of sorghum Caused by fungus *Sphacelotheda sorghi*. Smut affects the ovaries and all the ovaries in a head, as a rule infected.

Grain smut of sorghum (*Sorghum vulgare*) Caused by *Sphacelotheca sorghi*. All grains in the affected head transformed into elongated cylindrical sacs, consisting of black spore mass. For control, treat seeds with finely ground sulphur or Agrosan GN before sowing @ 4 gms/kg and 2 gms/kg respectively.

Gram blight The causal organism is *Ascochyta rabiei* and is sometimes has an epidemic form on gram (*Cicer arietnum*).

Gram positive Like fungi, bacteria can be stained. One of the principal stains used in bacteriology is the Gram's stain, which consists essentially in staining bacteria with one of the aniline dyes, usually gentian violet. These are bacteria which do not retain the colour and are called Gram-negative.

Grape powdery mildews The causal organism is *Uncinula necator.* In favourable conditions, it sometimes appears in an epidemic form.

Grassy shoot of sugarcane (*Saccharum officinarum*) Caused by virus. Excessive tillering, sprouting of lateral buds, poor cane formation particularly in cotton crop. Treat seed with hot water (52°C for one hour); use healthy seed.

Green ear disease of pearl millet (*Pennisetum typhoideum*) Caused by *Sclerospora graminicola.* Ear wholly or partially transformed into loose, head of small, twisted, green leaf-like structures; affected leaves become white and later brown; whitish downy growth of sporangia appears on lower surface of leaves. Grow resistant varieties; rogue out the affected plants, preferably when the plants are young.

Grey blight and fruit spot of guava (*Psidium guava*) Caused by *Pestolotia psidi.* Spots with brown rings and grey centres having black dots within occur on leaves and fruits; spray leaves fruits with 1 per cent Bordeaux mixture.

Grey blight of tea Caused by *Pestalotia theae* on tea (*Thea sinensis*).

Grey leaf spot of coconut (Cocos nucifera) Caused by *Pestalotia palmarum.* Small yellow spots formed on leaf which turn grey with small black dots in centre. Cultivation and judicious manuring can control the disease.

Grey mildew of cotton (*Gossypium* spp.) Caused by *Ramularia areola.* Whitish frosty growth seen mainly on under surface of leaves. Spray crop with 1 per cent Bordeaux mixture.

Ground nut rosette It was first reported from southern India in 1926 and is gradually increasing in importance. It is a virus disease. Attacked plants are stunted, their leaves turn creamy yellow and are reduced in size and yields are very much reduced.

Guignardia bidwilli *See* black rot of grape.

Guignardia camelliae* and *Laestadia theae *See* copper blight of tea.

Gummosis Caused by the fungus *Botrytis cinaria.* It is a serious disease of Citrus trees. Application of a paste or wash made of half kilogram of zinc sulphate, half kilogram of copper sulphate and one kilogram of lime in two gallons of water. Certain root stocks are resistant to this. Also caused by *Phytophthora* spp. and is called brown rot gummosis. *See* exudates and exudation.

Gummosis of sugarcane (*Saccharum officinarum*) Caused by *Xanthomonas vasculorum.* Pale yellow stripes are formed on leaves which later turn brown. Canes become thin and the internode length is shortened. Fibrovascular bundles become red. Use disease free seed. Destroy plant disease debris.

H

Haplobiontic The alternation of generation in fungi is haplobiontic like Algae. *See* Plant Science.

Haustoria (1) A special structure in the parasitic flowering plants, which they embed in the bark of the host species. (2) Fungi that

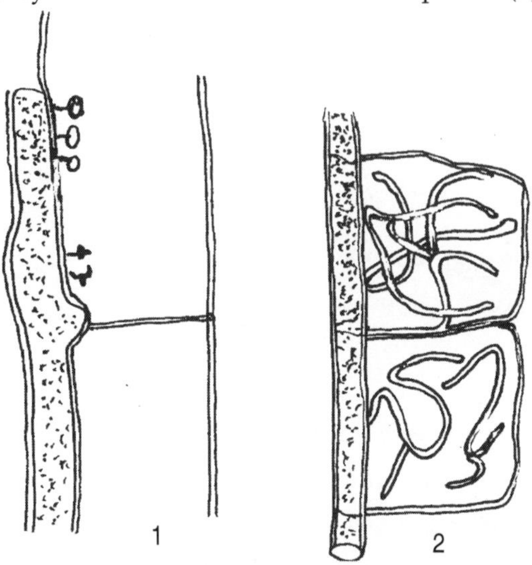

Haustoria

1: Cystopus; 2: Peronospora

penetrate the host and develop their mycelium within the tissues are known as endoparasites. The internal mycelium may be restricted to a small area as in leaf spot or it may be distributed throughout the plant, as in wilt diseases. It may be intercellular *i.e.* lying between the cells or intracellular *i.e.* lying with the cells. Both ectoparasites and endoparasites posses haustoria which are organs penetrating the cells and which have the function of absorbing food. They may be clavate, lobd, coraloid or filamentous.

Hayelis leaf blight of maize Caused by *Cochdiobolus heterostophus.* Also called *Drecholera maydis.*

Head scab of wheat Caused by *Fusarium nivele* and *F. graminacium.*

Head smut of sorghum (*Sorghum vulgare*) Caused by *Sphacelotheca reiliana.* All grains in infected head converted into elongated cylindrical sacs, consisting of black spore mass. Treat seed with Agrosan GN before sowing @ 2 gms/kg; practise crop rotation.

Head smut of Johnson grass Same as for sorghum.

Head smut of maize (*Zea mays*) Caused by *Spthacelotheca reiliana.* Smut sori replace tassel and ear forming hard, black, sooty mass; affected flowers grow into leafy structures; affected plants remain stunted, brittle and green longer than others. Crop rotation and sanitation is useful for its check.

Heart rot and internal black spot of beet (*Beta vulgaris*) Young unfolding leaves fail to develop normally, eventually turn black and die; dormant buds in aisles of older leaves grow out only to die. The disease is caused by deficiency of boron. Apply Borax (20–75 lbs/acre) to the soil and along with fertilizers.

Helicosporaeae A family of Moniliales (fungi imperfecti) in which spores are cylindrical, spirally twisted, one or more celled.

Helminthosporium avenae *See* leaf blotch of oats.

Helminthosporium frumentacei *See* leaf spot of barnyard millet.

Helminthosporium graminium *See* stripe of barley.

Helminthosporium halodes *See* leaf rot of coconut.

Helminthosporium leaf blight of finger millet (*Eleusine coracana*) Causal organism is *Helminthosporium leucostylum, H. nodulsum* and *H. tetramera*. Spots appear on both sides of leaf; later spots become enlarged and dark brown in colour; *H. nodulosum* causes stripes or shredding of leaves, seedling blight and head blight. For control, treat seeds with Agrosan GN before sowing @ 2 gms/kg seed.

Helminthosporium leucostylum Finger millet. *See* Helminthosporium leaf blight.

Helminthosporium oryzae It is a leaf spot disease of rice (*Oryza sativa*). It wide spreads all over the world.

Helminthosporium sacchari *See* leaf spot of sugarcane.

Helminthosporium sativum *See* root rot and foot rot of barley.

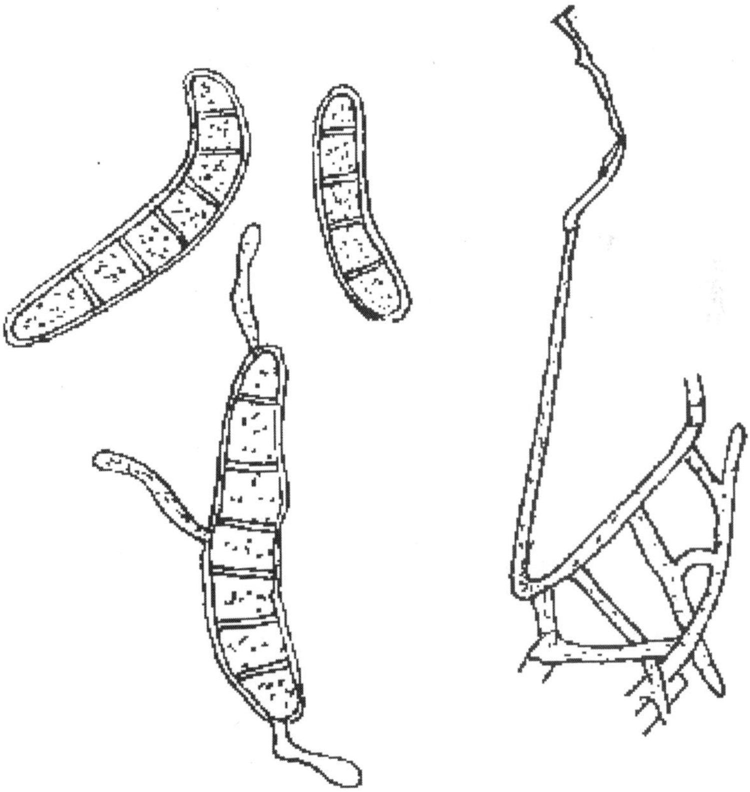

Helminthosporium oryzae

1: Conidia; 2: Conidophores

Helminthosporium sigmoideum* var. *Irregularae*–See* irregular stem rot of paddy.

Helminthosporium turcicum *See* leaf blight of maize.

Helotiales This order of Ascomycetes consists mostly of saprophytic fungi, but some species of the genera seudopeziza, Sclerotinia and Ciboria are parasites on economic plants.

Hemi Half.

Hemiascomycetes It is a sub-class of Ascomycetes in which asci are formed singly, directly from the zygote, sometimes closely aggregated, but no ascocarp formed.

Hemibasidiomycetes A sub-class of Basidiomycetes in which basidia arise from a single probasidium, septate or aseptate, often irregular, basidiospores irregular in number, usually forming secondary spores by budding.

Hemileia vastatrix It is the leaf rust of coffee. It is a major limiting factor in coffee production.

Hemileia This species of coffee must has been reported from Cameroons and it produces symptoms and morphology is also different than *Hemileia coffeicola*.

Hetero Different.

Heterodera rostochiensis Golden nematode of potato.

Heterodra avanae *See* Cereal root eelworm of wheat and barley.

Heterogamous Where male and female gametes differ in size and structure the process of fertilization is heterogamous.

Hetrothallic having two sexes on different hyphae *See progametangium*.

Hetrothallic Where the sexes occur in different thalli. Also called bisexual.

Heterotrophic bacteria A majority of bacteria are heterotrophic as they cannot' manufacture their food and thus make their food by the oxidation of inorganic compounds but obtain it from other sources like dead organic matter or living organisms.

Hill bunt of wheat Caused by *Tellitia caries*.

Histoplasmolysis *See* Animal Science.

History of biology *See* Plant Science.

Homiascomycetes Class Ascomycetes. This is a group of yeasts.

Horsegram leaf spot Caused by *Cercospora dolichi.*

Host An organism used as a source of nourishment by another organism, the parasite, which lives in or on the body of the host. In a definite host, the parasite reaches sexual maturity. In an intermediate host, the resting stage or young stage of the parasite are supported. *See* parasitism.

Hot water method This is a method devised for controlling loose smut as commonly practised. The seed is soaked in water at a temperature of 26–30°C for five hours. This induces the dormant mycelium to germinate. The seed is then quickly transferred to warmer water at 54°C for ten minutes, which helps killing the germ tubes. The thermal death point of the fungus and the seed, however, near each other, and the margin of safety is thus very narrow. Extreme care has thus to be taken in treating wheat seed against loose smut by this method.

Modification of this method has been devised by keeping the suspected diseased seed in hot sun, where temperatures in summers are usually very high. The seed is placed in shallow flat bottomed tubs, and covered with water so that the water level is about two inches above the level of grain. The tubs are placed in sum early in the morning and allowed to remain there for about five hours. At the end of the periods, the water is decanted and the soaked grain is spread on the threshing yard in the sum to dry.

Humaria granulata On this plant, the ascogonium retains its shape and forms an *ascogenous hyphae,* but the trichogyne disappears. Finally in *Humaria rutilans,* the ascogonium itself has disappeared but a fusion of vegetative cells occurs.

Hyalodictyae It is section of the Dictosporae (Monilialea), in which spores are muriform, ovate or elongate and hayline or bright.

Hyalodidymae It is a section of Dimerosporae (Moniales) in which spores are two celled, egg shaped or elongate, hayline or bright.

Hyalosporae This is a section of Amerosporae (Moniales) in which spores are one celled, round or egg shaped, elongate, bright or hayline.

Hydropericardium syndrome A disease of chicken. *See* Animal Science.

Hylophragmae This is a section of Phragsporae (Moniales) in which spores are three or more celled, oblong, hayline or bright.

Hymenium (1) A layer of the fruiting body of certain ascomycete and basidiomycete fungi in which the asci or basidia are borne. The hymenium may be directly exposed to the air as in the gills of the mushroom (a basidiomycete) and the apothecia of certain ascomycetes (*e.g.* the pezizales), or may open into a flask-shaped cavity, as in the perithecia of the ascomycete order sphaeriales. (2) The part of the sporophore which actually bears the spores in mushrooms is known as hymenium.

Hypae In fungi, a fine non-photosynthetic tubular filament that spreads to form a loose network termed as mycelium or aggregates into fruiting bodies (*e.g.* toadstools). Hyphae may be branched or unbranched and mayor many not have cross walls (septa) dividing them into cells. They are parasitic or saprophytic and the tips secrete enzymes to digest and penetrate the food supply. The hyphal walls of most species of fungi differ from those of plants in being composed of a nitrogenous compound called chitin or a form of fungal cellulose. Hyphae also differ in lacking plastids, any pigment being contained in the walls, cytoplasm, or oil globules. Hyphae are also found in the algae, in which they may form compact pseudo-parenchymatous tissues as in the thalloid seaweeds.

Hypnospores After successful dissemination, fungal spores may not find suitable conditions for growth in the new areas. In such cases, fungi form special resting spores or dormant structures known as hypnospores. *See* Palmella stage.

I

Imperfect stage of fungi *See* fungi imperfecti.

Importance of bacteria Bacteria and fungi, more perhaps than any other, determine the rate and character of organic synthesis and simplification. They govern by their enzymatic activity, the nature of humus and control, the volume of carbon dioxide produced.

Importance of fungi in soils Many kinds of mushrooms fungi producing such fruiting bodies as mushrooms, toadstools and puffballs are found in forests. The mycelia of certain members of this group sometimes infest the roots of trees giving an association called mycorrhiza. This is apparently of advantage to both the host and the parasite. In acid soils, mostly prevalent in forests, and in which nitrification is often negligible, the associated fungus may enable the trees to make use of complex nitrogenous forms that otherwise would remain unavailable.

Inclusion bodies Within the tissues of the host, virus diseases cause certain pathological changes, the most characteristic being the appearance of inclusion bodies, of which there are two kinds– The first are amoeba-like, amorphous bodies which are vacuolate arid bear a rather superficial resemblance to protoaoans. The second are crystalline in character, fragile and extremely variable in shape.

Incubation period The period that elapses between the settling of spores, bacteria etc. on the host and the appearance of the disease is known as incubation period.

Infection The establishment of a pathogen inside the host following entry in which a parasitic relationship between two organisms is established.

Infection-degree Symbols have been used as assigned to indicate degrees of infection. These symbols are–(-) Trace; (-) Slight; (-) Moderate; (-) Considerable; (-) Abundant; (;) Hypersensitive flecks and (.) Necrotic lesions.

Infection threads On germination, the ascospores from infection threads which fasten themselves to the leaf surface. The infections press against the thick cuticle, and the fungus slowly grows within it.

Infectious bursal disease It is a highly contagious acute disease in young chicken upto 6 weeks of age causing high morbidity and low mortality. Bursa may atrophy at the time of death. Sometimes streaky haemorrhages are also seen on the pelvic and pectoral muscles. Also known as 'Gomboro disease' it is caused by a virus of the family Birnaviridae which has two segments of double standard RNA genome. Natural infection provides a good immunity in infected birds. Cell-culture pasaged live vaccine and bursa-adapted inactivated vaccine are available.

Inflorescence rot of jack fruit (*Artocarpus integrefolai*) Young female inflorescence rots and sheds. The disease is caused by *Rhizopus artocarpus.*

Inoculum It is the infection material that can cause disease and it is that portion of individual pathogens that is brought into contact with the host.

Inoculum potential It is the degree of infectivity. The energy of a fungal parasite available for infection of a host at the surface of the host organs to be affected.

Irish blight of potato Same as late blight of potato.

Irpex destruens *See* Stump rot of Tea.

Irregular leaf spot of groundnut (*Arachis hypogea*) Caused by *Mycosphaerella arachidicola.* Affected leaves show irregular circular spots; become larger in size; bright yellow circular halo formed on upper surface of leaf. Spray with 1 per cent Bordeaux mixture; good cultivation and crop rotation; grow resistant varieties.

Irregular stem rot of paddy (*Oryza sativa*) Caused by *Helminthosporium sigmoideum* var. *irregularae.* Pale brown patches occur on lower internodes, diseased culms hollow; amber coloured and black sclerotia found on inner surface in dark grey weft of mycelium.

Isogamous If the male and female gametes are morphologically undifferentiated, this process of fertilization is isogamous.

J

Johne's disease It is a specific chronic contagious disease of cattle, sheep, goats, camels, buffaloes and occasionally of pigs. The disease is characterised by progressive emaciation and in cattle and buffaloes by chronic diarrhoea and thickening of the intestine. The presence of acid-fast organisms in the faeces samples confirms the diagnosis. The causal organism of this disease is *Mycobacterium paratuberculosis.* Under natural conditions, the disease spread by ingestion of feed and water contaminated by the faeces, of the infected animals. The animals aged from 3 to 6 years mostly suffer from this disease. The organism is more resistant to chemotheuropetic agents *in vitro* than *Mycobacterium tuberculosis.* The affected animals should be segregated and their faeces properly disposed of. A live vaccine consisting of a non-pathogenic strain has been developed to reduce incidence of clinical disease.

Jowar sugary disease caused exudation of sugary substances, the causal organism is *Sphacelia sorgii.*

K

Karnal bunt of wheat (*Triticum aestivum*) Caused by *Neovassia indica*. Diseased partly converted into black sooty powder which smells

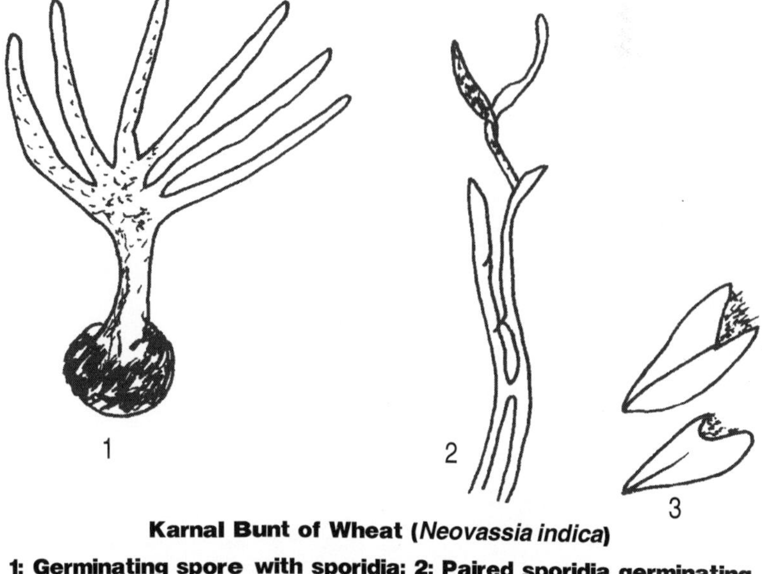

Karnal Bunt of Wheat (*Neovassia indica*)

1: Germinating spore with sporidia; 2: Paired sporidia germinating to form secondary sporidium; 3: Partially infected grains

like rotten fish; only a few years and a few grains in each ear affected. For control, grow resistant varieties.

Knapsack sprayers A machine used for spraying various materials on the crops, in which pumping is worked with one hand while the operator directs the spray with the other.

Koleroga disease of areca palms (*Areca catechu*) It is a rot disease, serious in areca palms in western peninsular India. The causal organism is *Phytophthora aracoe;* problem in growing crop.

L

Lagenidiales One of the orders of the sub-class Oomycetes.

Late blight *See* blight.

Late blight of potao (*Solanum tubersum*) caused by *Phytophthora infestens*. Brownish to purplish-black lesions on leaflets; petiole and stems; lesions enlarge rapidly; brown to purple discolouration of tuber followed by brownish dry rot; wet rot starts in storage. Spray foliage with 1 per cent Bordeaux mixture @ 100 galls/acre; remove foliage tops a few days in advance of harvest; avoid injury to tubers at harvest; sort out infected tubers.

Late blight of tomato (*Lycopersicum esculentum*) caused by *Phytophthora infestens*. Brown to purple black lesions of foliage, which advance rapidly under humid, warm weather; grey green water soaked spots on leaves, which enlarge to indefinite shape. Spray with 1 per cent Bordeaux mixture, Zenibs (Diathane Z78 and Parzate) or Nabam plus zinc sulphate.

Late wilt of maize Caused by *Cephalosporium maydis*.

Leaf blight of Arecanuts and beetlenuts (*Areca catechu*) caused by *Coniothyrium arecae*. Spots appear on leaves, which cause shedding. Spray leaves with Bordeaux mixture.

Leaf blight of durum wheat This disease is caused by *Fusariumnivele* in India.

Leaf blight of maize (Zea mays) caused by *Helminthosporium turcicum.* Long, elliptical, greenish-brown lesions on leaves; affected leaves thin and semitransparent; diseased plants look burnt or frost stricken. For control, practise crop rotation and sanitation; grow resistant varieties.

Leaf blight of maize caused by *Setospheriatiatoceca* or *Serobilum tritincum.*

Leaf blight of wheat (*Triticum aestivum*) *Alternaria triticina.* Whole or part of the leaf becomes blighted on account of brown oval spots. To control, grow resistant varieties.

Leaf blotch of oats (*Avena sativa*) caused by *Helminthosporium avenae.* Oblong and linear blotches with irregular margin on leaf blades; light reddish-brown in colour and frequently with sunken centres; infected leaves turn yellow and dry up. For control, treat seed with Cerasan or Agrosan GN before sowing @ 2.5 gms/kg seed.

Leaf blotch of pyrethrum (*Chrysanthemum cinarariaefolium*) caused by *Septoria chrysanthemella.* Brown leaf spots with. black centre cause leaf shedding. Treat seed with hot water (50°C for 30 minutes); spray crop with Bordeaux mixture.

Leaf blotch of turmeric (*Curcuma longa*) caused by *Taphrina maculans.* Blotches occur on leaves causing withering. Spray with 0.5 per cent Bordeaux mixture.

Leaf blotch of wheat (*Triticum aestivum*) caused by *Septoria tritice.* Light green to yellow blotches on leaves which later turn light brown, studded with black dots. Control measures are the same as for glume blotch.

Leaf curl of chillies (*Capsicum annuum*) caused by virus, smalling of leaves accompanied by downward curling. Give insecticidal spray.

Leaf curl of papaya caused by a virus and is a mosaic formed like on other plants also.

Leaf curl of tomato caused by virus. Dwarfing and puckering of leaves with mottling excessive branching, stunting of plants and partial or complete sterility. Controlling white fly vector *(Bemissia gossipiparida)* by insecticide *e.g.* Folidol or Ekatox and late transplanting may also help.

Leaf curl of peach and apricot Caused by *Taphrina deformans.* Curling puckering and distortion of young leaves which develop a red hue. Prune affected twigs; spray 1 per cent Bordeaux mixture before buds swell.

Leaf curl fungi These are facultative saprophytes.

Leaf curl of tobacco (*Nicotiana* spp.) Caused by virus. Curling, dwarfing and distortion of leaves. Rogue out affected plants; control vector with insecticides; weed out alternative hosts.

Leaf fall, fruit rot and bark canker of citrus Caused by *Phytophthora palmivora.* Water-soaked lesions on leaves, green twigs and fruits of mandarin; bark canker later appears on smaller branches; crown rot and girdling of stem occurs. Prune dead twigs before monsoon and spray with Bordeaux mixture (5–5–50). Repeat spraying after two months.

Leaf mould of tomato (*Lycopersicum esculentum*) Caused by *Clasdosporium fulvum.* Chlorotic spots with indefinite margin appear on upper side of leaflet; velvety olive green mould on lower side; when many spots merge, entire leaf droops and dies; flower and fruits also infected; yellow sunken spots appear in ripening fruits. Spray with 1 per cent Bordeaux mixture.

Leaf roll It is one of the five virus diseased found in potato.

Leaf roll of potato Caused by a virus. Upward rolling of margins, of leaflets; tissue assumes rigid leathery texture; whole top becomes more erect, followed by necrosis; of phloem growth retarded and tubers become undersized. Use disease free certified seed; spray crop with insecticide to control aphids.

Leaf rot of coconut (*Cocos nucifera*) Causal organism is *Helminthosporium halodes.* Distal ends of leaflets blacken and shrivel; redish brown lesions on tender leaflets; soft rot develops on central shoot. Spray crop with 1 per cent Bordeaux mixture; better cultivation and liberal application of potash.

Leaf rust It is also called brown or orange leaf rust of wheat and barley and is due to *Puccinia triticina.*

Leaf rust of barley Caused by *Puccinia hordei.* Small round, light yellowish-brown pustules on leaves; later pustules black, covered by epidermis. For control, grown resistant varieties.

Leaf rust of maize (*Zea mays*) Caused by *Puccinia sorghii.* Reddish brown open pustules on leaves which sometimes dry up. For control, grow resistant varieties and practise crop rotation.

Leaf rust of wheat (*Tritivum aestivum*) Caused by *Puccinia triticina.* Round or oblong, orange pustules on leaf sheath; pustules small and irregularly arranged; later pustules turn black and remain covered with a thin membrane. Controlled by growing resistant varieties.

Leaf smut of paddy (*Oryza sativa*) Caused by *Entyloma oryzae.* Small angular or elongate, black spots appear scattered on leaves below epidermis.

Leaf spots These are the more important and familiar among the symptoms produced by fungi; brown being the characteristic colour of dead tissue; it is the colour of many leaf-spots. They may be minute or several millimetres in diameter.

Leaf spot of beet (*Beta vulgaris*) Small circular spots occur in large numbers caused by the fungus *Cercospora beticol.* Centres of the leaf spot turns brown; whole leaf becomes senescent and dies prematurely.

Leaf spot and blight of *Brassica* spp. Caused by *Alternaria brassiciola* and *A. brassicae.* Small dark coloured spots, which spread rapidly to form circular lessions; in humid weather, concentric rings of dark conidiophores appear; cauliflower develops brown discolouration; cabbage after harvest acquires black mouldy appearance. Hot water treatment of seed at 50°C for 30 minutes. Spray crop with 1 per cent Bordeaux mixture at pod stage.

Leaf spot of brinjal (*Solanum melongena*) Caused by *Cercospora melongena.* Chlorotic spots turning grey to brown with age; when spots are numerous leaf shedding takes place. Spray with 1 per cent Bordeaux mixture.

Leaf spot of broad bean (*Vicia faba*) Caused by *Alternaria brassicae.* *See* under mustard.

Leaf spot of castor (*Ricinus communis*) Caused by *Cercospora ricinella.* Minute black or brown spots with pale green margins on both surfaces of leaves; later on colour of spots turns pale brown and then greyish-white with deep brown margins; black tiny dot-like structures also formed in the centre of these spots. For control, grow resistant varieties.

Leaf spot of chillies (*Capsicum annuum*) Caused by *Cercospora capsci.* Small circular water soaked spots appear on leaves, stem, petiole and peduncle of fruits; centres of old spots droop out; whole leaf turns yellow and drops and falls off. Spray with Bordeaux mixture 1 per cent solution.

Leaf spot of cluster beans (*Cyamopsis tetragonoloba*) Caused by *Alternaria brassicae.* Same as in mustard.

Leaf spot of cotton (*Gossypium* spp.) Caused by *Mycosphaerella gossypina.* Round irregular, dark brown spots with ashy centre formed on leaves, which later coalesce; whole leaf dries up. Use resistant varieties.

Leaf spot of cowpea (*Vigna sinensis*) Causal organism is *Cercospora canescens* and *C. cruenta.* Same as for *Phaseolus* spp.

Leaf spot of fenugreek (*Trigonella foenum-graecum*) Caused by *Cercospora traversiana.* Brown, elliptic or round spots with dark margin appear on leaves; later, dark olive growth appears in centre of spot. Grow resistant varieties; spray with 1 per cent Bordeaux mixture.

Leaf spot of lablab bean (*Dolichos lablab*) Caused by *Cercospora dolichi.* Circular to angular spots with grey centre and reddish border appearing on leaves. Spray with 1 per cent Bordeaux mixtrure.

Leaf spot of lady's finger (*Abelmoschos esculentus*) Caused by *Cercospora hibisi.* Brown spots on leaves. Spray with 1 per cent Bordeaux mixture.

Leaf spot of lima bean (*Phaseolus lanatus*) Caused by *Cercospora cruenta.* Circular to angular spots, grey in centre and often surrounded by reddish borders, appear on older leaves; pods also affected. Spray crop with 1 per cent Bordeaux mixture.

Leaf spot of barnyard millet (*Echinochloa frumentacea*) Caused by *Helminthosporium frumentacacei.* Numerous yellow, dot-like spots appear on both sides of leaves and sheaths; gradually spots increase lengthwise and coalesce form long stripes; older spots give a dark brown or chocolate appearance.

Leaf spot (Sigatoka) of banana Caused by *Mycospherela musicora* and *Cercospora musae.* Leaf spots formed. Apply an oil based spray with a low volume sprayer.

Leaf spot of maize (*Zea mays*) Caused by *Cercospora sorghii.* Circular to somewhat irregular brown spots formed on the leaves which later become olivaceous.

Leaf spot of pearl millet (*Pennisetum typhoideum*) Caused by *Curvularia penniseti.* Affected leaves show dirty brown spots mostly on edges of leaves, which coalesce to form irregular patches; leaf sheaths and ears also attacked.

Leaf spot of pigeon pea (*Cajanus cajan*) Causal organism is *Cercospora indica.* Small, light brown, interveinal leaf spots with dark brown centre, coalescing to form bigger spots; later leaves dry, curl and fall off; lesions on petioles and stem; profuse dark olivaceous mycelial growth on under surface of leaf during wet season. For control, practise crop sanitation; spray with Bordeaux mixture.

Leaf spot of rocket salad (*Euraca sativa*) Caused by *Alternaria brassicae.* Same as for mustard.

Leaf spot of safflower (*Carthamus tinctiorium*) Caused by fungus *Cercospora carthami.* Small circular or irregular, slightly sunken dark brown spots having yellowish tinge on the border appear on leaves; affected leaves turn brown, get distorted and form holes; sometimes stems and floral parts also attacked. Spray with Bordeaux mixture for control.

Leaf spot of sesamum (*Sesamum orientale*) Caused by *Cercospora sesami.* Numerous golden yellow spots found on leaves and pods. Treat seeds with hot water before sowing.

Leaf spot of sesbania Caused by *Cercospora sesbaniae.* Small, yellow, irregular spots are seen on upper surface of leaves; lower surface show scant, effuse dark brown mycelial growth. Destroy the affected plant debris.

Leaf spot of Soya bean (*Glycine max*) Caused by *Cercospora sojina.* Numerous typical zonate grey spots with purplish-brown margins appearing on leaves and other parts of plant except roots. Remove all the diseased plant debris.

Leaf spot of sunflower (*Helianthus annuus*) Caused by *Alternaria tenuis.* Small, oval, depressed whitish spots on leaves.

Leaf spot of sweet clover (*Melilotus* spp.) *Cercospora devisii.* Same as for Fenugreek.

Leaf spot of sweet potato (*Ipomea batatas*) Caused by *Cercospora batatae* and *C. bataticola.* Large brown spots on leaves; severe damage to leaves in warm moist weather. Spray with 1 per cent Bordeaux mixture.

Leaf spot of sword bean (*Canavalia ensiformis*) Caused by *Cercospora canavaliae.* Leaf spot not distinct; black effuse patches; appear on lower surface of leaf. *See* as in *Phaseolus* spp.

Leaf spot of turmeric (*Curcuma longa*) Caused by *Colletotrichum capsici.* Oblong brown spots, with black concentric rings within, occur on leaves. Spray with Bordeaux mixture (3-3-50).

Leaf spot of velvet bean (*Mucuna* sp.) Caused by *Cercospora stizolobii.* Circular or confluent or irregular spots; tan to brown in colour, found mainly on leaves; margins raised, and dark brown to back in colour. Method of control is same as for *Paseolus* spp.

Leaky fruit rot of brinjal (*Solanum melongena*) Caused by *Pythium aphanidermatum.* Infection occurs on fruit at blossom end in touch with the soil; turns tissue watery; covered by cottony mycelium. Discard all discoloured fruits at harvest.

Leleillula taurica *See* powdery mildew of cotton.

Leptomitales One of the orders of the sub-class Oomycetes of Phycomycetes.

Leptosphearia sacchari Commonly known as the ring spot disease of sugarcane *(Saccharum officinarum).*

Leptosphearia salvinii It is the perfect state of *Sclerotium oryzae.* It belongs to Ascomycetes.

Leptospirosis *See* Animal Science.

Lesion A pathogenic fungus when attacks its host, more or less rapidly kills the tissues. This killing is known as necrosis. The diseased part of the.plant where such necrosis has taken place is known as a lesion.

Leveillula The erysiphe mycelium is superficial in this.

Leveillula laurica *See* powdery mildew of castor, pigeon pea and cotton.

Life cycle of loose smut of wheat

Lime sulphur *See* sulphur fungicides.

Linseed rust caused by *Melampsora lini* in which the teliospores germinate only when they are subjected to freezing temperature for long periods.

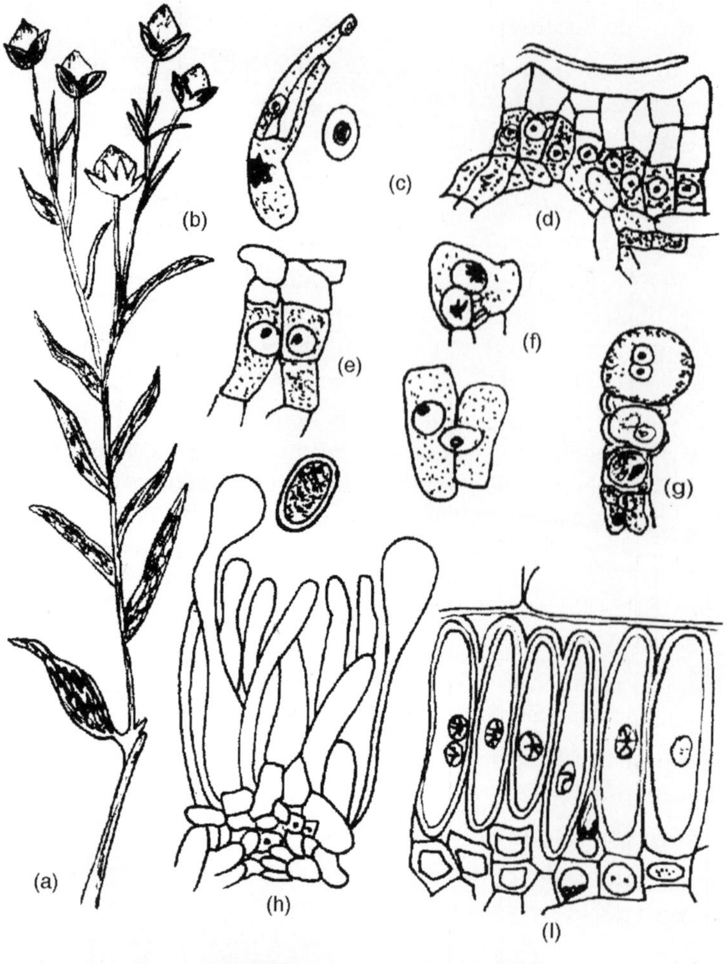

Linseed Rust (*Melampsora lini*)

(a) Symptoms; (b) Pycniophore; (c) Pycniospore; (d) Aecium; (e) Paired cells at the base of aecium; (f) Fusion of cells; (g) A chain of binucleate aeciospores in formation; (h) Uredium and uredospores (section); (i) Telium and teliospores (section)

Listeriosis A bacterial disease of domestic and wild animals. It is caused by *Listeria monocytogenes* a gram positive small non-sporing coccobacillus seen in soil, silage, sewage and faeces of domestic and wild animals and birds. To control, avoid potential source of infection. *See* Animal Science.

Little leaf of brinjal (*Solanum melongena*) Caused by virus. Affected plants produce tiny yellow leaves and does not bear brinjals. Destruction of alternate hosts like Datura; periodical dusting of plants with BHC to control the vector.

Little rod It is the meaning of bacteria in Greek. It was coined by a German Prof. F. J. Cohn in the middle of the nineteenth century. These are unicellular plants and along with all other unicellular plants and animals and viruses are known as micro-organism.

Localised mycelium In such cases, the mycelium extends to a short distance into the host tissue near the place of infection.

Long-cycle rust If all the spore forms are present in a rust, then it is known as an euform or macrocyclic or long-cycled rust.

Long smut of sorghum (*Sorghum vulgare*) Caused by *Tolyposporium ehrenbergii*. Only a few grains in a panicle attacked and converted into long, cylindrical and slightly curved bodies which release brownish green spore mass on rupturing. For control, grow resistant varieties.

Loose smut of barley (*Hordeum vulgare*) Causal organism is *Ustilago nuda*. Smutted heads; grains replaced by black powdery mass of spores; later leaving the rachis naked. Solar heat or hot water treatment of seeds; grow resistant varieties for control.

Loose smut of Johnson grass (*Sorghum halepense*) Caused by *Sphaceletheca cruenta*. Same as for Sorghums.

Loose smut of oats *Ustilago avenae* is the causal organism.

Loose smut of wheat (*Triticum aestivum*) Caused by *Ustilago tritici*. Same as of barely.

Loose smut of sorghum (Jwar *Sorghum vulgare*) Causal organism is *Sphacelotheca cruenta*. All spikelets smutted; those which escape turn into leafy structures; affected grains replaced by loose powdery black spore mass. For control, treat seed with Agrosan GN before sowing @ 2 gms/kg.

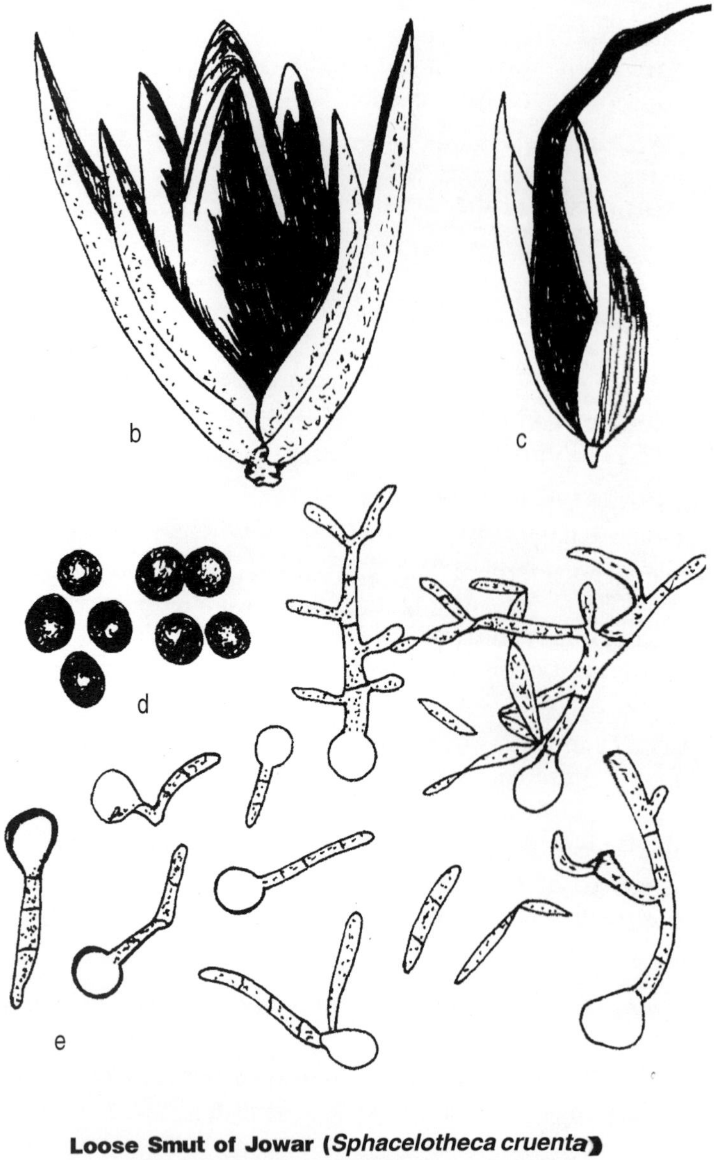

Loose Smut of Jowar (*Sphacelotheca cruenta*)

(a) Smutted ear; (b) Sores involving stamens;

(c) Exposed columella after the spores have fallen;

(d) and (e) Spores and their germination

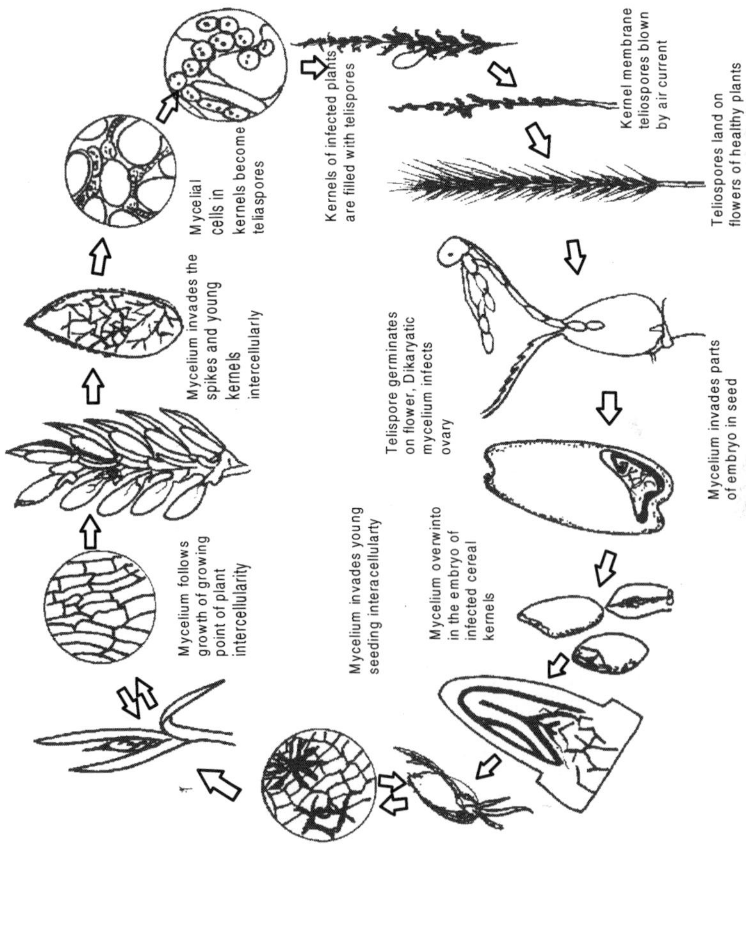

Life Cycle of Loose Smut of Wheat (*Ustilago tritici*)

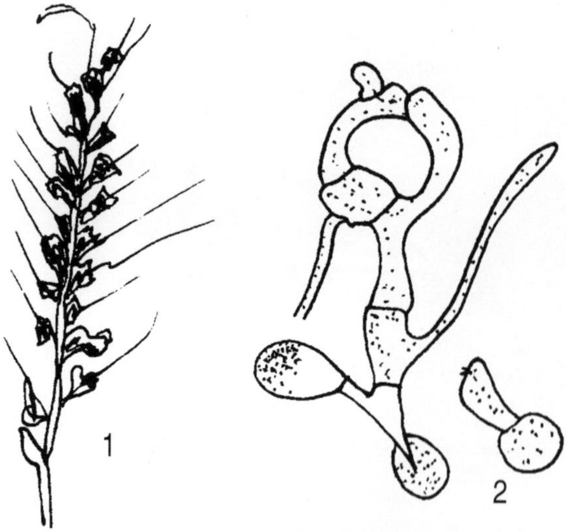

Loose Smut of Wheat

1: Wheat ears affected by smut,
2: Germination of spores of ustilago tritici

Lomasome An infolding of the plasma membrane found particularly in fungal hyphae and spores, and also in some algae and higher plants. Lomasomes are similar to the mesosomes of bacteria and possibly have equivalent functions.

Lumping ill *See* Animal Science.

M

Macroconidia In fungi imperfecti, infections take place through the tender lateral roots and rootless and possibly root-hairs, but at what precise age of the plant it occurs has not yet been determined. That it never takes place in the parts above ground is quite certain. Mycelium is hayline and produces spores of three types within the tissues of the host plant-macroconidia, microconidia and chlamydospores. Macroconidia are produced in small cushions of stromatic mycelium on the surface of the bark. They are long curved, pointed at the ends septate and measure 15-50μ in length and 3-5μ in breadth.

Macro-cyclic rusts *See* long cycled rusts.

Macrophoma musae *See* black finger of banana and plantain.

Macrophomina Genera of Sphaeropsidales, have many members which cause plant disease of economic plants such as *M. phaseoli*.

Macrophomina phaseoli Attacks cotton, sesamum and soyabean. Also attack horsegram and lentil. *See* charcoal rot of finger millet, dry root-rot of gram and cotton. Also *See* for cowpea and cotton.

***Macrophomina phaseoli* of Jute** *See* stem rot or dry rot of jute.

Mahonia A genus whose species serve as a alternate host for rusts.

Main stalk rot A serious disease of banana is caused by a fungus *Gloeosporium* spp. and *Botryodiplodia* spp.

Main stock and finger stalk rot of banana and plantain Caused by *Botryodiplodia* spp. and *Gloeosporum* spp. Rotting of the main stalk of the bunch and of stalks of fingers. Cut the main stalk well above the top most hand; apply some dust fungicide to the cut end of the stalk and tie a piece of grease proof paper over it; cure the bunch at a comparatively lower temperature (20–25°C).

Maize leaf spot Caused by *Phacosphorasis maydis.*

Mango mildew *Oidium mangiferae.*

Marek disease of poultry *See* Animal Science.

Melampsora lini *See* rust of linseed.

Melampsora ricini *See* rust of castor.

Melanconiales One of the orders of fungi imperfecti having conidia borne in definite acervuli and finally free in the substratum.

Melenopsichium eleusinis *See* smut of finger millet.

Meliola butleri *See* sooty mould of Citrus.

Merk's disease *See* Animal Science.

Metula The conidia in *Aspergillus repens* are multinucleate while generally in other *Aspergillus* spp., they are uninucleate. In some species, the phialides are not produced directly on the vesicle but through small cells known as metula.

Microbes All the unicellular plants and animals, bacteria and viruses are known as micro-organisms or simply microbes.

Micro-teliospores Aecedoid–Teliospores.

Micorrohizal fungi Fungi that exist in association with roots of plants.

Micrococus amylovorous A bacterium causing fire blight roots of plants.

Microconidia These are small elliptical or curved, unicellular or with one or two septa, and measure 5–15µ in length and 2–4µ in breadth. They are formed within the tissues and usually become free on falling, but in culture media and on the surface of the plant, they are held together in a little drop of liquid at the tip of the hypha, until as many as dozen or more are present in the drop.

Miceocyclic rusts *See* long cycle rusts.

Mildew (downy) of grape *See* downy mildew.

Mildew (downy) of peas *See* downy mildew. Caused by *Erysiphe polygoni.* Always appears in an epidemic form every year. It has been shown to have specialised races, each restricted to a particular host or group of related hosts. The form on garden peas *(Pisum sativum)* can attack field peas *(Pisum arvens)* but not other hosts.

Midew of peach, apricot (*Prunus persica, Prunus armeniaca*) Caused by *Sphaerotheca pannosa.* Tiny white superficial spots on leaves which become powdery as they enlarge. Dust with sulphur powder.

Mildew (powdery) of barley (*Hordeum vulgare*) Caused by *Erysiphe graminis.*

Minus strain In heterothallic mycelia with different sexual tendencies are morphologically indistinguishable, though a slight difference in the luxuriance of their growth can sometimes be distinguished. One strain is designated as plus (+) and the other minus (–).

Molya A local name of the disease caused by *Heterodera avenae* of the nematodes.

Moniliales These fungi are also known as *Hyphomycetes,* have a septate mycelium which is branched, and hayline or dark coloured. Reproduction is closely by means of conidia, which are produced in the most varied manner on the conidiophore or their branches, conidia have many different shapes and colours. Conidiophores may be short, upright branches or bound into coremia or in layer-like bolsters. Some of the more important genera whose members cause diseases in plants are Alternaria, Fusarium, Cercospora, Helminthosporium and Piricularia.

Monoblepharidaes An order of sub-class Oomycetes.

Mosaic of apples Caused by virus. Mottling of the leaves with small irregular creamy white or yellow spots which may coalesce, and form large chlorotic tissue develop among some of the larger veins. The mottling may take the form of a light and dark green patches. Avoid use of scion from infected plants.

Mosaic of banana and plantain Caused by virus. Light and dark green patches formed on leaves. Control same as for bunchy top.

Mosaic of cardamom (*Elettaria cardamomum*) Caused by virus. Dark and light green streaks formed in the spindle; clump declines and dies. Eradication of diseased clumps; propagation of cardamom from seed.

Mosaic of chillies (*Capsicum annuum*) Caused by virus. Light and dark green patches observed on leaves. Rogue out infected plants; grow resistant varieties.

Mosaic of cowpea (*Vigna sinensis*) Caused by virus. Yellow and green patches appear on leaves; mottling very conspicuously; affected plants remain stunted. For control, rogue out the affected plants in early stage of the disease.

Mosaic of cucurbits Caused by virus. Mosaic on young leaves near growing tip; V shaped necrotic areas appear on older leaves, extending from margin to mid-rib; dwarfing of internodes; young fruits show mild mottling to extensive watery malformation.

Mosaic disease Diseases of plants caused by viruses.

Mosaic of figs Caused by virus. Mosaic mottling of leaves of affected plants.

Mosaic of groundnut (*Arachis hypogea*) Caused by virus. Mottling of leaves, accompanied by malformation, smalling and puckering.

Mosaic of peas Caused by virus. Clearing of veins followed by chlorosis or severe yellowing of leaves with numerous dark green areas dispersed over the leaflets. There is general stunting of the infected plants. Rogue out diseased plants; use insecticide to control insect vector; grow resistant varieties.

Mosaic of potato (*Solanum tubersum*) Caused by virus. A number of viruses are responsible for producing different symptoms on different varieties of potatoes. Some viruses *e.g.* virus X are transmitted mechanically while others *e.g.* virus A and Y are transmitted both mechanically as well as through insect vectors. Plant disease free seed or certified seed.

Mosaic of sann-hemp (*Crotolaria juncea*) Caused by virus. Leaves of affected plants show mosaic mottling.

Mosaic of sugarcane Yellowish green mottling caused by virus. Grow resistant varieties.

Mosaic of tobacco (*Nicotiana* spp.) Caused by virus. Mottling of leaves in dark and light green streak; virus mechanically transmitted. Rogue out diseased plants; strick observance of sanitary measures.

Mosaic of tomato Caused by virus. Several types of mosaic attack tomato such as tobacco virus. Avoid infected soil and tobacco cultivation; disinfect seed-beds; avoid contamination of hands with virus, and wash hands clean with soap and water. Cucumber mosaic is filiform, shoe string like filiage; pronounced dwarfing of plants. Potato mosaic shows vein clearing and mottling; combination of cucumber and potato mosaic cause severe necrotic disease known as winter streak or double streak.

Motile spores They are endogenously produced and the cell within which they are fashioned is known as sporangium.

Mottle leaf Of Citrus. Also called Frenching or foliocellosis. Could be controlled by injecting or spraying of Ferrous sulphate or Ferrous phosphate. Also caused by zinc deficiency. Internal spaces on leaves of sweet orange, pomelo and lemon turns yellow; leaves become small; growth arrested. Spray young flush with zinc sulphate lime mixture (10–5–100); apply green leaf manure.

Mottling In leaves. A condition when leaves' surface become uneven and rough.

Mottling of leaves Caused by mosaic disease. Transmitted through insect vectors.

Mucor A fungus commonly known as black mould. It includes about 50 species distributed all over the world. *See* zysporangium. Commonly known as bread mould or black mould. It grows saprophytically in a variety of habitats like moist bread, jam, leather, paper, pickle and other foodstuffs. Some species are parasitic and cause soft rot disease of plants. Other parasitic species within the tissue of various animals, particularly in lungs causing a disease known as lung mycosis. It has some positive importance. It is used in alcoholic fermentation of rice by using *M. javanicus* and *M. praini.* Citric acid is also prepared from molasses with the help of *Mucor* spp.

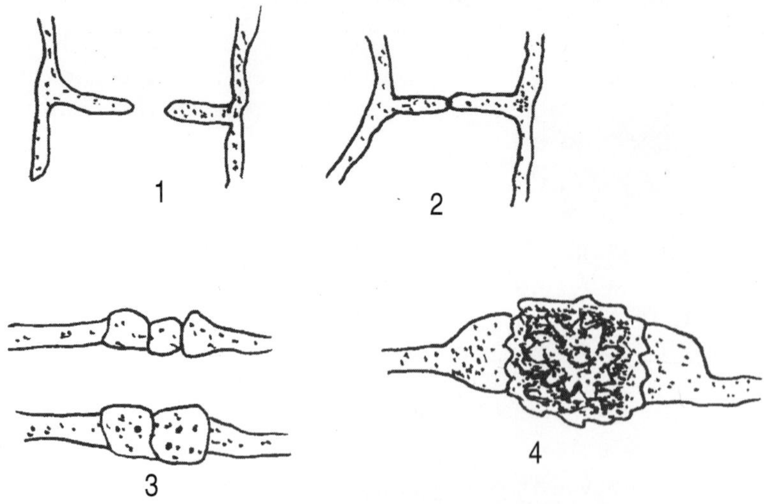

Mucor mucedo
Stages in the formation of zygospores

Mushrooms Belong to the sub-class Eubasidiomycetes, characterised by a true basidium. The field mushrooms *Psalliota compestris* is the best known fungus and it is a highly organised fungus in the Eumycetes.

Mustard The major diseases of this crop are *Peronospora brassicae,* a downy mildew of *Brassica compostris* (mustard). Other important disease is the *Cystopus candidus,* the white rust of this crop.

Mycelia sterli An order of fungi imperfect and consists of a few species that cannot be accommodated in any of the other orders, for they do not have even the conidial fructifications are placed in this order.

Mycelium (1) A filamentous grass comprising the body of a fungus, each filament being called a hypha; the mycelium often forms a loose mesh as in Mucor, but the hyphae may become organised into definite structures *e.g.* the fruiting body of a mushroom. The mycelium produces the reproductive organs of a fungus. The whole thallus of unicellular fungi may be thus employed,

Coenocytic mycelium

but only part of the thallus produces gametangia or sporangia in most species, the rest of the thallus being vegetative. *See* hypha. (2) The vegetative body of a fungus consists of a network of richly branched filaments known as hyphae, and a mass of hyphae is termed mycelium. Hyphae grow by terminal and apical elongations. They are segmented into cells by cross walls which are called septa and segmented mycelium is known as septate. It is frequently perennial in the substratum which may be wood, roots and stems of living trees, humus.

Mycetozoa The controversial Myxomycetes are sometimes called Mycetozoa.

Mycology Fungi are thallophytic plants of very varied forms and habits and comprise a large and heterogeneous group. The fungi are delimited from the animal kingdom by the presence of a cell-wall. The science which deals with their life histories, relationships and evolutionary process is known as Mycology.

Mycoplasmosis *See* Animal Science.

Mycoplasmosis synoviae A disease of poultry. *See* Animal Science.

Mycorrhiza In case of certain gymnosperms, absorption is helped by certain fungi in association with roots. This association is mycorrhiza.

Mycoses A few members of Aspergillaceae are parasitic to animals and cause the disease known as mycoses.

Mycosphaerella arachidicola *See* irregular leaf spot of groundnut.

Mycosphaerella gossypina *See* leaf spot of cotton.

Mycosphaerella personata *See* Tikka disease of groundnut.

Mycosphaerella musicola *See* leaf spot of banana.

Mycosphaerella rabici *See* blight of gram crop.

Mycotoxicosis A disease of animals and birds. *See* Animal Science.

Myxobacteria (Slime bacteria) Saprophytic gliding bacteria that may produce brightly coloured fruiting bodies visible to the naked eye. The cells, which occur in masses of slime, are flexible unicellular rods that divide by transverse fission. Myxobacteria are common in soil, animal dung, and decaying plant matter.

Myxomycetes The largest class of slime fungi, which have variously been placed in both the Myxomycophya and the Eumycophyta. The fungal body is a plasmodium, and ingests food, but reproduction is by flagellate spores formed in sporangia. Also *See* Mycetozoa.

Myxomycophyta (Slime fungi) A division of simple organisms showing both plant and animal characteristics. They are multinucleate, lack cell walls, and display amoboeid movement ingesting food, leading some to group them with the Protozoa as Mycetozoa. However, in some genera spores, with cellulose cell walls are formed in sporangia; this being a plant characteristic. The Myxomycophyta includes the classes Acrasiales and Labryinthulales; in its wider sense it also includes the Myxomycetes and the Plasmodiophorales. *See* also plasmodium.

N

Narrow brown leaf of paddy (*Oryza sativa*) Caused by *Cercospora oryzae*. Light brown lesions with dark centre on leaves and leaf sheaths; lesions on peduncles narrow; those on glumes short and laterally spreading. For control, grow resistant varieties.

Nature of soil organisms causing diseases of plants It is plant forms of soil life, in general, exert the most devastating effect on higher plants. All three groups–bacteria, fungi and actionmyces–contribute the role of plant diseases.

Nature of virus disease *See* Animal Science.

Necator This is a stage in the disease *Corticium salmonicolor*, called pink disease of organs. This is a pustural stage known as necator stage. The pustules of this stage are in vertically elongated streaks and generally orange red.

Necrosis When a pathogenic fungus enters and attacks its host, it starts killing of plant tissues.

Nectria cinnabarina *See* Dieback of Tea.

Nematocyst Same as thread cell.

Nematoda A large phylum of marine, freshwater and terrestrial invertebrates, the roundworms. Most are free-living *e.g.* *Anguillula* (vinegar eel), but many are parasites *e.g.* *Heterodera* (eelworm of potatoes) and *Ascaris* (found in pigs and human intestines). Some cause serious diseases in man *e.g. Wuchereria* (causing elephantiasis). Nematodes are bilaterally symmetrical with an unsegmented smooth cylindrical body pointed at both ends and covered with a tough cuticle.

The body cavity is not a true coelom and there are no blood or respiratory systems. The muscular and excretory systems and embryonic development are unusual. Nematodes are not closely related to any other phylum.

A Parasitic Nematode

Neovossia horrida *See* bunt of paddy.

Neovossia indica A causal organism of the disease Kamal bunt of wheat.

Nephotettix apicalis Stunt of rice is transmitted by the leaf hopper, but other leaf hoppers that though feed on the plants are not able to transmit the disease.

Net blotch of barley (*Hordeum vulgare*) Caused by *Helminthosporium teres.* Scattered oblong, dark brown blotches on leaves; not extended to leaf sheath; lesions also on floral bracts. For control, same as for covered smut.

Nocardiosis *See* Animal Science.

Nursery damping off and collar rot of coffee (*Coffee arabica*) caused by *Pellicularia filamentosa (Rhizoctania solani)*. Stem gets constricted, rots extends down in the root. Reduce shade and restrict watering; drench seed bed with Perenox (2 ozs in 2 galls per 24 sq.ft.); far yellowing cuproxide (1.5 ozs in 2 galls); Fermate (1 oz in 2 galls), or Aresan (2/3 ozs. in 2 galls).

O

Obligate aerobes Those bacteria which require free molecular oxygen for their respiration are called obligate aerobes. Some bacteria respire in absence of oxygen and for them oxygen is toxic. These are named as obligate anaerobic.

Obligate parasites Parasites which are restricted to living tissues (also known as biotrophs) are the obligate parasites.

Oidia (1) These are a kind of vegetatively formed cells which arise from the breaking up of an aerial hyphae into a number of equal, rounded segments, functioning also as resting bodies. But oidia do not have a resting periods, and germinate as soon as suitable conditions are available. (2) In Mucor (Fungi), the mycelium of M. racemosa when immersed in a strong sugar solution and is poorly supplied with oxygen becomes segmented into short chain of cells. These cells round off remain thin walled and are called oidia.

Oidiopsis In Erysiphales, sometimes the mycelium is entirely superficial except in Leveillulá, and its conidial stage oidiopsis, where it is endophytic. It is hayline, branched and septate and its cells are uninucleate.

Oidium mangifera Causes shedding of inflorescence of mango trees.

Oidium tingitanium Due to this fungus, sometimes the leaves turn yellow and etiolated, as in citrus seedlings attacked by this. Also *See* powdery mildew of Citrus.

Oidium erysiphoides A disease of sann-hemp powdery mildew.

***Odium* spp.** Powdery mildew of jute.

Oldenlandia aspera This plant is attacked by the rust *Coleosporium oldenlandae*. It is the synonym of old name *Uredo oldenlandae.*

Olpiopdiosis While the gametes are motile isogametes in *Synchytrium endobioticum*, they are non-motile heterogametes in Olpidiopsis.

Oogamous In *Albugo* sp., the sexual reproduction is oogamous. Towards the end of the growing period of the host, some of the hyphae penetrate deeper into the tissue of the host and form sex organs. Due to their hidden position were disovered later.

Oogamous reproduction In Phytophthora, sexual reproduction is oogamous. Oogonia are spherical to pyriform, smooth and hayline to yellowish. Inside they have an oosphere which is free and which after fertilisation is generally amphigynous and the antheridial attachment persists below the oogonium.

Oogonia The female gametangia are know as oogonia or ascogonia.

Oomycetes One of the two sub-classes of Phycomycetes. In this, the gametangia are dissimilar; perfect state represented by the oospores; imperfect state by zoosporangia germinating by the production of zoospores or less commonly functioning directly as conidia.

Ooplasm During the development of oogonium in Fungi (Albugo), in the beginning the cytoplasm is evenly vacuolated and the nuclei are uniformly distributed. Soon it differentiates into an outer vacuolated periplasm and the inner denser ooplasm (oosphere).

Oospores *See* Oomycetes. In one or two species of Cystopus, more than one nucleus remain functional and are fertilized by the entry of an equal number of male nuclei. The resulting oospores are thus mulinucleate and are known as compound oospores. Parathenogenetic spores are also known to occur.

Opercullela padwickii *See* root-rot of gram.

Operculum In some cases, the asci are provided with a lid-like structure for the ejection of spores–the operculum.

Opsis forms Some rusts produce pycnia, aecia and telia, the urediia being eliminated from their life-cycle. The oraeciospores or teliospores then assume the role of summer or distributing spores. They are known as opsisspores.

Orange leaf rust *Puccina triticina.*

Organo-mercurials Organo-mercurials have proved very effective for the control of seed-borne and some soil-borne diseases. They are applied as dusts in extremely small dosages to the seed before sowing. Cereal and millet seeds may be treated in this way as a regular practice. They have not harmful effect on germination, prevent emergence losses caused by soil-borne fungi and give a better stand. There are a number of organo-mercury preparations of which Agrosan GN, a proprietary product, is one. They are, however, poisonous to mean and beasts, hence great care is necessary in handling these fungicides and the seeds treated with them. Some vegetable seeds and other like linseed are non-tolerant of mercury. For them, non poisonous organic fungicides like Nomersan, a proprietary product, are available.

Organo-sulphur powder *See* sulphur fungicides.

Orobanche cernua Fam. Orobanchaceae. An annual parasite on root of crops like tobacco and *Pennisetum typhoides.* Eradication by chemicals without affecting the host plant is difficult for removal of immature plant as it emerges from soil and before it develops seeds prevents its spread. Heavily infested areas to be reclaimed by growing a trap crop and spraying with 1–2 lb 2, 4–0. *See* also brown rape of tobacco and brinjal.

Oosphere In the higher Phycomycetes where copulation between antheridia and oogonia takes place directly, the oogonium consists of homogenous protoplasm to begin with, but this later contracts to form a large spherical naked body within the oogonium, which is known as the oosphere.

Ostiole Pycnidia are provided with openings at their apical end, known as ostiole.

Over-summering Vegetative mycelium is of short duration. After a time, it forms over wintering or over-summering structures.

These structures can withstand heat, cold and draught, and tide over adverse conditions.

Ozonium wilt of cluster beans (*Cymopsis tetragonoloba*) Caused by *Ozonium texanum var. parasiticum.* Rotting of collar region, white weft of mycelium formed on soil near base of plant, studded with white to light brown, round sclerotia; diseased plants wilt and dry. Crop rotation cultural practices can improve the control of this fungus.

P

Palm leaf pustules A serious disease of date palm leaves caused by *Graphiola phonicis.*

Pan sukh disease of paddy (*Oryza sativa*) Affected plants dry up; root development and tillering poor; head, if produced are light, flowers mostly sterile. In case of early disease, drain off the field, dry it and then irrigate again; when disease appears late, apply ammonium sulphate at 30–50 lbs N per acre.

Panama disease Of banana. Also called banana wilt disease. It is caused by the fungus *Fusarium cubens,* sometimes it acquires such serious situation that the whole plantation has to be abandoned in the affected areas.

Papaya This plant suffers from many destructive diseases such as leaf curl, stem rot etc.

Paraphyses Sterile hyphae.

Parasites These fungi obtain their food from the living tissues of plants or animals they parasitize on, and they are therefore harmful organisms.

Parasites and parasitic diseases In case of such diseases, it is always the parasite that derives the benefits and the host usually suffers from various diseases and injuries in varying degrees.

Parasites races Parasitic specialization presumes that these are, within well-defined morphological species of fungi, entities which are not easily distinguishable by morphological characters but which differ from each other in their physiological and parasitic characters.

Parasitic specialization Only a totality of individuals with a morphological resemblance, among themselves, and their progeny, but possesses essentially similar physiological and parasitic qualities. The recognition of this fact has led to the discovery of the phenomenon of parasitic specialization. As fungi respond differentially to environmental conditions, their plasticity has to be recognised, a species therefore represents not only a totality of individuals with a morphological resemblance among themselves and their progeny, but possesses essentially similar physiological and parasitic qualities. The recognition of these qualities has led to the discovery of the phenomenon of parasitic specialization, and understanding of which is highly desirable.

Parasitic varieties These are actually classified into two categories: varieties and races. Parasitic varieties are distinguished to a small extent by their morphology, but chiefly by their capacity to infect certain species of one or more genera of the same family, which may, however, be attacked by another parasitic variety of the same fungus.

Parasitism An association between two organisms in which one, the parasite, benefits at the expense of the other, the host. The tolerance of the host varies from being almost unaffected to serious illness and often death. An obligate parasite can live in association with a host, whereas a facultative parasite can exist in other ways, for example as a saprophyte.

Pasteurollosis *See* Animal Science.

Pathogenic bacteria *See* Animal Science.

Pathogenic fungi *See* Animal Science.

Pathogenecity In conducting pathogenic tests, a few precautions have to be taken. The tests should be conducted in greenhouses if they are available. The variety of the host plant must be, as a rule, the same as the one on which the disease had first appeared. Temperature and humidity conditions within greenhouse must also be such as would favour the appearance of the disease. Bright and direct sunlight must be avoided.

Pathogenesis The sequence of progress in disease development from the initial contact between a pathogen and its host to the completion of syndrome.

Pathology Study dealing with the diseases of, their causes, symptoms and their control, cure and elimination.

Pathotype Subdivision of species distinguished by the common characters of pathogenecity, particularly in relation to the range of hosts.

Patwa *See* stem rot.

Peliospore The morphology of teliospore furnishes the principal characteristics for the differences of genera of the Uredinales. Teliospores are dormant spores and germinate as a rule after a period of rest by the formation of long promycelium which is usually four-celled. Each of the promycelial cells is provided with a haploid nucleus formed as a result of the divisions of the fusion nucleus, the first being a reduction division. A single sporidium is formed on each of the promycelial cells on a short pedicel-like structure, the strigamata. Several divisions from this common method have been observed. The exceptions are– teliospores of Catenulospora germinate without rest periods; those Melompsora are sessile Teliospores of Coleosporium form an internal promycelium *i.e.* the teliospore becomes four-celled each being a promycelial cell.

Pellicularia Several species of this genus cause disease of various plants of economic importance, *P. koleroga, P. filamentosa,* formerly known as *Corticium solani* attack many plants groups. *P. koleroga* is an important rot-disease of coffee while *P. filamentosa* affects potatoes, chillies and other crops. Its Rhizoctonia stage (known as *R. solani,* is well known parasite of the seed-beds and of cotton in which it causes a troublesome root-rot.

Pellicularia filamentosa *See* root-rot of cow pea.

Pellicularia (Sclerotium) rolfsii *See* stem rot of tobacco.

Pellicularia salmonicolor *See* pink disease of tea.

Penicillin It is an important curative drug for animals.

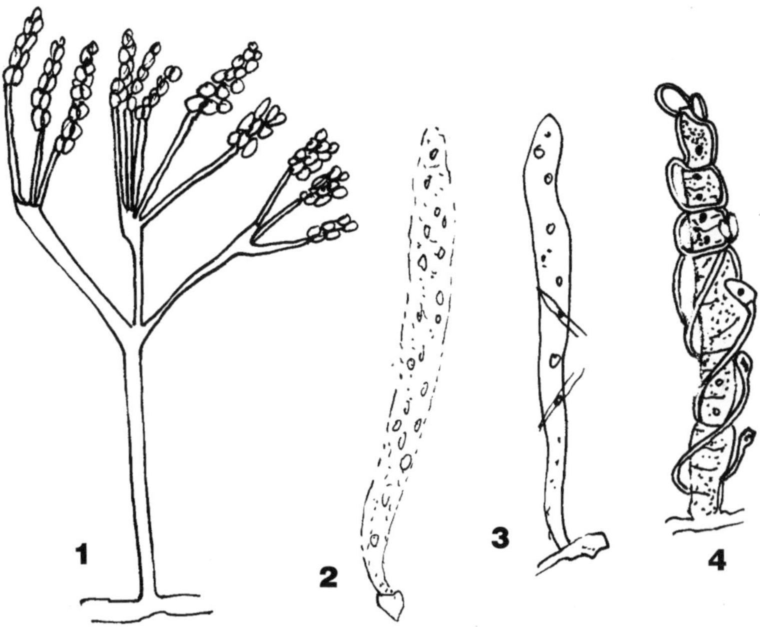

Pencillum verviculatum

1: Condidiosphore, sterigemata (phialides) and conidia;
2: Multinucleate ascogonium and young antheridium;
3: Fusion of antheridium and ascogonium;
4: Ascogonium surrounded by sterile hyphae

Penicillium verviculatum It is heterothallic, the ascogonium when young is uninucleate but divides and re-divides to form thirty-two or sixty-four daughter nuclei.

Pentalonia nigronervosa It is vector of the virus disease that attacks cardomoms. This vector also transmits the bunchy-top disease of bananas.

Peridium Soon after the plasmogamy and the formation of the asci, the stalk cells of both the sexes organs branch freely to form a large number of sterile hyphae which coil themselves around the archicarp to form a distinct sheath called peridium. In Puccinia, the aecium is usually bell-shaped and enclosed by a one-celled wall, known as peridium. In Ascomycetes, perithicia are round or flask shaped ascocarps. An outside wall known as peridium encloses the hymenium.

The genus has been established for the temporary accommodation of such species whose teletial stages are unknown, but as soon as they are discovered, the species are transferred to the right genus.

Periphyses Asci form a dense hymenium or they may be more loosely arranged. Delicate and evanescent paraphyses are present, but paraphyses *i.e.* paraphyseslide threads at the edge of the hymenium (but not intermingled with the asci) are more frequent.

Perenospora viticola Causes downy mildew disease of grapes.

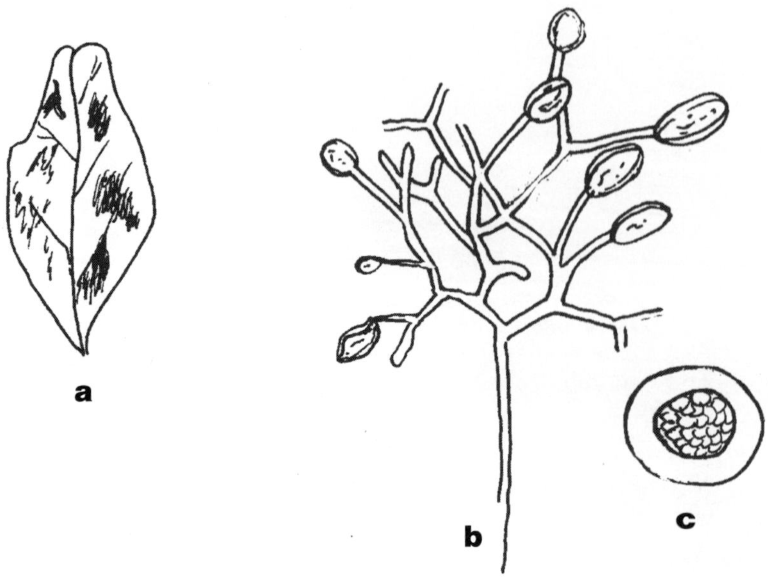

***Perenospora pisi* on Pea Leaf**
a: Symptoms, b: Conidiophore and conidia, c: Oospore

Perenox It is a good soil fungicide against Pythium disease.

Perfect state The perfect state of *Sclerotium rolfsi,* which is known as *Corticium rolfsii,* occurs only when the dishes in which the fungus is growing are exposed to a higher temperature of 35–40°C.

Perithecium A kind of ascocarp of flask shape.

Peronospora This disease causes the disease of downy mildews. In this, the conidiophores are dichomatously branched, conidia germinate from the sides.

Peronospora arborescens *See* down mildew of poppy.

Peronospora aestivalis *See* down mildew of mustard crop.

Peronospora brassicae This causes atrophy in Brassica where there is a suppression of the floral bud formation.

Peronospora lathyripalistris *See* downy mildew of Lathyrus.

Peronospora pisi causes downy mildew of peas.

Peronospora trigonellae *See* down mildew of fenugreek.

Peronospora parasitica Causes mildew in Brassicas. Critical studies have shown that this is a composite species and that a large number of smaller morphological and biological species which are usually specialised on single host genera and often on single species host, can be segregated.

Pestolotia An important genus of Melanconiales *P. theae* causes grey blight in tea. This is the commonest blight of this crop.

Pestolotia psidi *See* grey blight and fruit spot of guava.

Pestal otiapalmarum *See* grey leaf spot of coconuts.

Phenological relations In diseases caused by *Cercospora arachidicola* and *Mycosporella arachidicola,* if such relations are precisely determined, and the time of the appearance of fungus noted, then only a successful spray schedule can be made.

Phialopore *See* plakae.

Phillipine downy mildew of maize Caused by *Peronosclerospora phillipansis.*

Phomales Same as Phyllotistictales or Sphaeropsidales.

Phomopsis The genus belongs to Sphaeropsicales of Ascomycetes. It is an important genus causing destructive disease of plants.

Phomopsis blight and fruit rot of brinjal (*Solanum melongena*) Caused by *Phomopsis vexans.* Dark brown lesions on the stem. Round to oval spots are formed on leaves. Diseased fruits show soft and watery lesions which later become black and mumified.

Black dot like pycnidia are observed on the diseased portions. Use resistant varieties.

Phialide During development of conidia in *Aspergillus niger* (Fungus) some of the vegetative cells of the mycelium, arise erect, thick walled, unbranched, non-septate hyphae known as conidiophores. The cell from which a conidiophore is formed is the foot-cell. The tip of the conidiophore becomes swollen to form a club-shaped or rounded structure called vesicle. From the surface of the vesicle arise a large number of buds called phialides (sterigmata). In some species, the phialides are not produced directly on the vesicle but through small cells called matula.

Phragmosporae A family of fungi imperfecti identified as having spores three more celled, oblong or fusion.

Phycomyces nitens In this heterothallic species, single zygospore may produce both (+) and (–) aplanospores and even a few spores that are homothalic in their sexual tendencies.

Phycomycetes A class of true fungi including many important parasites (*e.g.* Pythium) and moulds (*e.g.* Mucor). Many are found in damp or aquatic habitats and there are few fully terrestrial species. Sexual reproduction is either oogamous or by conjugation. The Phycomycetes are divided into subclasses according to the number of flagella borne on the asexual spores. These are the lowest of class of fungi, and include aquatic and land forms. In the lower, the mycelium is poorly developed; but in the higher forms, it shows better development. Hyphae never form strands or tissues and the thallus ranges in structure from isolated cells in the lower families to a richly branched mycelium in the families higher up in the class. These are divided into two subclasses as Oomycetes and Zygomycetes.

Phyllody of *Sesamum orientale* Caused by virus. Floral parts of affected plants transformed into green leafy structures followed by profuse branching; later, affected plants assume a branchy top; no seed is formed.

Phyllosticta A genus of sphaeropsidales.

Phyllostictales *See* Phomales.

Physalospora mutila Its imperfect state is *Sphaeropsis malorum*.

Physalospora tucumanensis This is supposed to be the ascigenous stage of *Colletotrichum falcatum.*

Physiological specialization The existence of physiologically distinct but morphologically identical races within a species. Such physiological races are important in host-pathogen studies, particularly in planning programmes to breed for crop resistance. The cereal rust fungus *Puccinia graminis,* for instance, has over 200 physiological races, which have developed in response to new cereal cultivars as they come on the market.

Physiotype A population of pathogen in which all individuals have a particular character of physiology but not pathogenecity in common.

Physiological Yellowing disease of sugarcane. It is a mineral deficiency disease of the crop.

Physiologic race Pathogens of the same species and variety which are morphologically similar but which differ in pathological characteristics.

Physiologic races impact Hybridization between different races in case of black rust of wheat in the aerialhost, giving rise to new physiological races. Physiologic races are common among *Phytophthora infestens, Puccinia glumarum, Puccinia graminis, Puccinia triticina, Uromyces appendiculatus* and *Ustilago tritici.*

Physoderma zeamaydis *See* brown spot of maize.

Physoderma alfalfae *See* crown wart of alfalfa.

Phythium stock rot of maize Caused by *P. aphanidermatum.*

Phytomonas citri It is a short, rod shaped organism, motile by means of a single polar flagellum. It is enclosed in a capsule and is strictly aerobic. *See* canker of citrus.

Phytomonas malvacearum This fungus causes angular leaf-spot of cotton caused by bacteria.

Phytomonas solanacearum This is a vascular disease caused by bacteria. They are present in abundance in the lumen of xylem vessels in the early stage of the disease. This clogging interferes with the supply of water to the foliage, and the plants wilt in consequence.

Phytomonas tritici It is a motile rod with one *Pokar flagellum.* Its thermal death point is 50°C.

Phytomonas tumefaciens This bacterium causes a hyperplastic disease by formation of tubercles, tumours or galls as in crown gall of fruit-trees.

Phytophthora cambivora* and *Rhizoctonia solani *See* wilt of pyrethrum.

Phytophthora colocasiae This is colocasia blight organism and occurs in rainy season only. Plants recover as soon as dry conditions return.

Phytophthora infestens *See* late blight of potato.

Phytophthora nicotinana In black shank of tobacco plants, this is the causal organism, attacks the foot of the host and brings about a sudden death.

Phytophthora palmivora *See* fruit rot of jack fruit; *See* black leg of tobacco; and also see leaf fall, fruit and bark cancer of citrus.

Phytophthora parasitica *See* seedling blight of castor; *See* wet rot and damping off of cotton *Phytophthora parntica;* and also *See* bean rot.

***Phytophthora* spp.** On Citrus. The production of gum on the lower part of the trunk and sometimes on the main roots. The most common forms are caused by members of the genus Phytophthora and are called brown rot gummosis, because the same fungi cause a rot of the fruit characterised by a pale brown colour and a distinctive smell. Where infection has taken place, treatment is possible if undertaken in time. The bark is scraped to find out how much is the disease. The wound is covered with a fungicide to prevent the re-infection. A paste or wash made of ½ kg of zinc sulphate, ½kg of copper sulphate, 1 kg of lime to 2 gallons of water. It has also been reported that carbolic acid diluted with an equal amount of water, but preferably creosote oil effective. These should be used with some kind of sticker.

Pigeon leaf stem canker This is caused on *Cajanus cajan* and the causal organism is *Diplodia cajan.* It attacks especially the collar region. The disease appears when the plants are two to three months old, in the form of grey, scutiform lesions with dark edges, on the stems, branches and twigs.

Pigeon pea wilt The causal organism is *Fusarium udum.* No other disease does more damage to pigeon peas than wilt does. Use resistant varieties.

Pilliculariafilamentosa *See* root rot of gram and groundnut.

Pillicularia koleroga *See* black rot of coffee.

Phillicularia rolfsii *See* scleorital disease of wheat and barley.

Pine apple disease of sugarcane (*Saccharum officinarum*) Caused by *Certaostomella paradoxa.* Outer tissue below the rind blight red, central core a sooty black mass of cottony tissue. Distinct odour of pineapple. Fungicidal seed treatment of setts.

Pink disease of banana (*Musa* spp.) and of apples Caused by *Corticium salmonicolor.* Same as for Tacti fruit. Pink disease of Cinchonoa. *See* in tea, caused by *Corticium salmonicolor.*

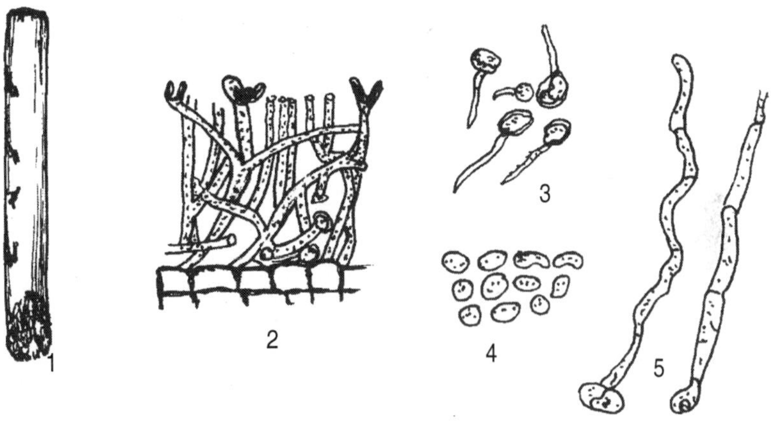

Pink Diseases

1. A twig with incrustation; 2: Section of sporophore showing hymenium; 3: Germinating basidiospores; 4: Nector spores; 5: Nector spores germinating

Pink disease of Citrus Caused by *Corticium salmonicolor.* This is a disease which attacks a great variety of citrus crops in the tropics, causing severe damage and frequently death. Its name comes from the pink mycelium on the surface of the branches at one stage. Search for cankers in the dry season and scraping them and the bark at the crotches where dormant mycelium has been observed and treating the wound with Bordeaux mixture paste or creosote. Frequent examination and the removal of diseased portions in the wet season is also necessary.

Pink disease of coffee (*Coffee arabica*) Caused by *Pillicularia salmonicolor*. *See* as in tea.

Pink disease of custard apple Caused by *Corticium* in *Anona squamosa*. *See* as in tea.

Pink disease of jack fruit (*Artocarpus integrefolia*) Caused by *Corticium salmonicolor*. Young woody branches lose their leaves and die back; thin pink incrustation appears on lower shaded side. Prune and burn affected branches; protect cut-surface with Bordeaux paste.

Pink disease of mango (*Mangifera indica*) Caused by *Corticium salmonicolor*. Same as in tea.

Pink disease of oranges (*Citrus aurantium*) *See* Pellicularia.

Pink disease of rubber Caused by *Pellicularia salmonicolor*.

Pink disease of tea (*Camellia sinensis*) Caused by *Pellicularia salmonicolor*. Young woody branches lose their leaves and die back; a film of silky threads over runs the branch towards the close of monsoon; a pink incrustation travels down the branch on lower side. Prune off and burn affected branches; spray spruned bushes and those surrounding with Bordeaux mixture.

Piricularia oryzae *See* Blast disease of paddy. This is also rotten-neck disease of paddy.

Pisum arvens *See* Erysiphe.

Pisum sativum *See* Erysiphe polygom.

Plant disease Plant disease may be defined as a condition in which the plant as a whole or any part of it does not perform its normal function. It may be brought about by micro-organism like fungi, bacteria, viruses, etc. or by physiological causes such as high or low temperature, lack or excess of soil moisture and aeration deficiency or excess of plant nutrients, soil acidity or alkalinity etc. The micro-organisms obtain their food either by breaking down dead plant and animal remains (saprophytes) or by attacking living plants and animals (parasites). In order to obtain their food nutrients, the parasitic organisms kill the tissue-cells of the host plant as a result of which either the whole plant or a part of it is damaged and killed, or considerable disturbance takes place in the normal metabolic processes of the living plant. In some cases, they also produce toxic substances.

Plant disease's control Micro-organisms are reproduced and disseminated mainly by spores. Wind, water, seed, diseased plants, cuttings and tubers, animals, men and insects, and soil help in their dissemination and spread of disease. Virus diseases are spread by direct contact, through cuttings and grafts, and more commonly by insect vectors. Most fungi and bacteria erupt from the surface to produce spores after the host tissues have been damaged. Virus works entirely internal. Hence, the control of plant diseases usually lies in preventing infection, as it is very difficult to kill the fungi or bacteria or to inactivate the virus inside the host plant. The evolution of disease resistant varieties of most of the economic crops has, however, revolutionised the method of disease control. A number of varieties resistant to certain important diseases have been evolved and are available for the use of farmers though it is very cheap method of plant disease control, it has its own limitations. Many chemicals are also used in control of plant diseases.

Plasmogamy Fusion of protoplasm, usually referring to the fusion of cytoplasm but not nuclei. Plasmogamy in the absence of karyogamy (fusion of nuclei) occurs between fungal mycelia of different strains to form a heterokaryon.

Plasmopara A genus of peronosporaceae in which both direct and indirect germination of conidia is replaced by the spongiophores.

Plasmodiophorales One of the eight orders of sub-class Oomycetes.

Plasmodiophora brassiceae *See* club rot of Crucifers.

Plasmopara viticola Now known by this name of the past name of *Pernospora viticola*.

Plasticity of fungi As fungi respond differentially to environmental conditions, their plasticity has to be recognised, and a species therefore represents not only the totality of individuals with a morphological resemblance among themselves and their progeny, but possesses essentially similar physiological and parasitic qualities.

Plectenchyma Hyphae may intertwine and form pod-like, thick tissues, called plectenchyma.

Pleocyta sacchari *See* rind disease of sugarcane.

Pleomorphic *See* as polymorphic.

Pleuropneumonia *See* Animal Science.

Pollinodium The branch of the hypha that produces the male sex organ (Aspergillus).

Podosphaera leucotricha *See* powdery mildew of apples.

Polymorphic or pleomorphic Fungi with more than one form of spores.

Polymorphism The tendency of a fungus to produce different kinds of spores and this is one of the devices fungi adopt to make their survival more secure.

Polynepharatis *See* Animal Science.

Porcine rotavirus *See* Animal Science.

Potato early blight *See Alternaria solani.*

Potato leaf-roll It is a virus disease of wide occurrence.

Potato late blight *See Phytophthora infestens.*

Potato 'rind-disease' This is caused by *Pathogenic bacteria.*

Potato virus disease Five potato virus diseases are identified *viz.* leaf-roll, X, Y, A and D; but other also await precise determination.

Powdery mildew of apple (*Pyrus malus*) Caused by *Podosphaera leucotricha.* White powdery growth on leaves, twigs, blossoms and fruits; deformation of infected structures. Spray (*i*) before the blossoms open, (*ii*) as the last petals fall, and (*iii*) two weeks later, with lime-sulphur (1–50) or ammonical copper carbonate (3–5–50).

Powdery mildew of barley (*Hordeum vulgare*) Caused by *Erysiphe graminis.* Light grey cushion like, cottony colonies with black specks found on above ground parts only. For control, dust with finely powdered sulphur @ 15–20 lbs/acre; grow resistant varieties.

Powdery mildew of lablab bean (*Dolichos bean*) Caused by *Leveillule taurica* var. *macrospora.* Grey white fungus on lower surface of leaves. Same as for powdery mildew.

Powdery mildew of broad bean (***Vicia faba***) Same as for Phaseolus. Causal organism is *Erysiphe polygoni.*

Powdery mildew of castor (*Ricinud communis*) Caused by the fungus *Levecillula tauricia. See* as in cotton.

Powdery mildew Caused by the species of *Erysiphecceae* and some others.

Powdery mildew of citrus Caused by *Oidium tingitanium.* Whitish mealy growth on young foliage and green twigs; leaves distort and shed; young fruits drop prematurely. Dust affected parts with sulphur or spray with wettable sulphur or Burgundy mixture (5–6¼–50).

Powdery mildew of coriander (*Coriandrum sativum*) Caused by *Erysiphe polygoni. See* as under double bean.

Powdery mildew of cotton (*Gossypium* spp.) Caused by *Leveillula taurica* and *Oidium* spp. White shiny powdery patches appear on under surface of leaves; corresponding of leaves becomes pale and brown. Dust crop with finely powdered sulphur to control the disease @ 15 lbs/acre.

Powdery mildew of cowpea (*Vigna sinensis*) Caused by *Erysiphe polygoni.* Same as for *Phaseolus* spp.

Powdery mildew of cucurbits Caused by *Erysiphe cichoracearum* and *Spharotheca humuli* var. *fuliginea.* Tiny white superficial spots appear on leaves. Dust crop with sulphur @ 15 lbs/acre.

Powdery mildew of cumin (*Cuminum cyminum*) Caused by *Erysiphe polygoni.* Same as under double bean.

Powdery mildew of fenugreek (*Trigonella foenum-graecm*) Same as for *Phaseolus* spp., caused by *Erysiphe polygoni.*

Powdery mildew of jute (*Chorchorus* spp.) Caused by *Oidium* spp. Same as in cotton.

Powdery mildew of lady's finger (*Abelmosches esculantus*) Caused by *Erysiphe cichoracearum.* Greyish mealy growth on under surface of leaves. Dust crop with sulphur @ 15 lbs/acre.

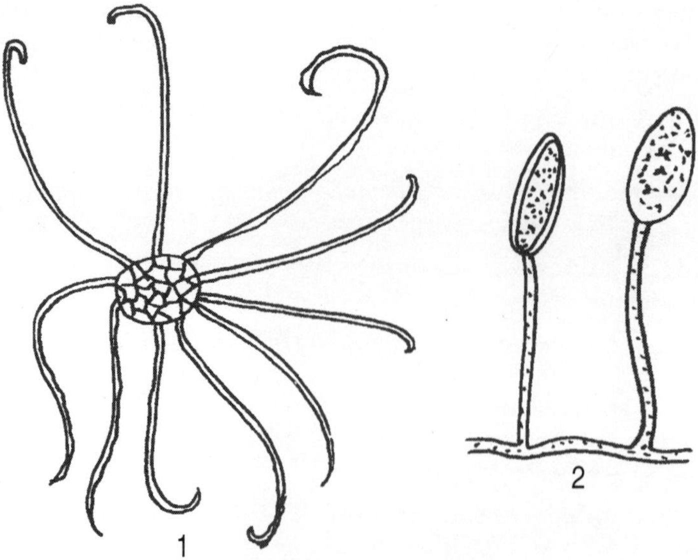

Powdery Mildew of Grape
1. Clestothecium of the same (uncinala necdator),
2. Conidiophores and conidia

Powdery mildew of lathyrus Same as *Phaseolus* spp.

Powdery mildew of lime bean (*Phaseolus lanatus*) Caused by *Erysiphe polygoni.* Faint slightly discoloured specks from which white powdery spots spread and over run leaf, stem and pods. Dust with finely powdered sulphur @ 15 lbs per acre.

Powdery mildew Of mango–*Erysiphe* spp. Of grapes–*Unicnulaxe necator.* Three dustings with sulphur control the disease. Of wheat–*Erysiphe graminium.*

Powdery mildew of mango (*Mangifera indica*) Caused by *Oidium mangifera.* Mealy white growth appears on inflorescence and tiny fruits causing their shedding. Dust inflorescence with sulphur once before flowers open and again when fruits are set.

Powdery mildew of peas (*Pisum sativum*) Caused by *Erysiphe polygoni.* Same as for beans.

Powdery mildew of pigeon pea (*Cajanas cajan*) Caused by *Leveillula taurica.* Same as in cotton.

Powdery mildew of *Phaseolus* spp. Caused by *Erysiphe polygoni.* White powdery patches on leaves and other green parts which later become dull coloured and studded with black dot-like structures. Grow resistant varieties; dust the crop with finely powdered sulphur (22 mesh).

Powdery mildew of poppy (*Papaver somniferum*) Caused by *Erysiphe polygoni.*

Powdery mildew of sann-hemp (*Crotolaria juncea*) The disease is caused by *Oidium erysiphoides.* Same as in cotton.

Powdery mildew of sweet clover (*Melilotus* spp.) Caused by *Erysiphe polygoni.* Same as for alfalfa.

Powdery mildew of tobacco (*Nicotiana* spp.) Caused by *Erysiphe cichoracearum.* Affected leaves appear dusted with white powder in patches; lower leaves first to be attacked. Spacing, clean cultivation and early priming off leaves; apply sulphur dust to soil around plants, or spray in the early growing stages with 1 per cent Bordeaux mixture.

Powdery mildew of tomato (*Lycopersicum esculentum*) Caused by *Leveillula taurica.* White mealy growth on leaves. Dust with sulphur @ 15 lbs/acre.

Powdery mildew of wheat (*Triticum aestivum*) Caused by *Erysiphe graminis.* Same as of barley.

Powdery scab of potato (*Solanum tubersum*) Caused by *Spongospora subterranea.* Faintly brown raised areas, ½ cm in size; diseased tissue become jelly like containing mature spore balls; epidermis rifts and peels black to expose spore balls; sori on tubers enlarge to form cankers; dry rot causes in storage. Practise crop rotation; spray crop with carbamates (*e.g.* Nabam plus Zinc sulphate, Dithane Z-78 and Parzate).

Predisposing factors Prevalence of a susceptible host is a condition for the occurrence of epiphytotics. The requisite host must not only be present, but it must also be susceptible and predisposed to attack. When a plant is rendered more susceptible to a parasite, plants of the same variety normally are, then it is said to be predisposed. Pre-treatment and disease control–A layer of the powder in case of formalin is diluted with an equal quantity of water and sprayed over a thin layer of seed. The seed is then heaped and kept covered with canvas overnight. For internally seed borne diseases, hot water treatment and solar heat treatment are very efficacious. For hot water treatment of loose smut of wheat, the seed is soaked in water for four to five hours and the dipped in hot water at 120°F (49°C) for ten minutes, and then immersed again in hot water at 129°F (54°C) for ten minutes. *See* hot water method.

Primary inoculum Over-wintering or over-summering stage may be in or on the seed of the host, in perennial wild, host, in debris of infected plant of the previous season, in dormant spores, sclerotia or other organs of the soil, or in the form of an inhabitant of the soil flora. Whatever, the source, it causes primary infection.

Principal organic mercury compounds Used as fungicides, are ethyl mercury phosphate, ethyl mercury chloride and chlorophenol mercury.

Progametangia Sexual reproduction in Phycomycetes if the isogamous type. Both the mating organs are alike. The protuberances arise one each one two neighbouring branches. These grow towards each other as a result of contact for a chemical stimulus. They grow in size and become the copulating branches known as progametangia.

Progametangium formation In sexual reproduction in heterothallic fungi (Mucor), two hyphae with different factors (+ and –) come in contact with each other, they form side branches at the point of contact. The growth of these branches at the point of contact and towards each other is controlled by volatile substances (hormones) which they secrete. Each branch swells at the tip, becomes club-shaped and is known as progametangium. Soon a transverse septum appears near the tip of the progametangium and divides it into a terminal part, the gametangium and a basal part the suspensor. The protoplast of each gametangium represents a coenogamete.

Promycelium *See* basidium.

Protoaeceia These are the haploid rudiments of the aerialfructifications.

Probasidium *See* Chlamydospore.

Protobasidiomycetes In these, asidia arise from a stalked probasidium of strictly limited growth, septate and four-celled, each bearing only one basidospore.

Pseudo-stem rot Of banana. Newly planted suckers turn black. It is advised to remove the pseudo stems before planting and dip the rhizome in some disinfectant before planting.

Pseudomonas Polar flagellate bacteria.

Pseudomonas rubrilineans *See* red stripe of sugarcane.

Pseudomonas solanacearum *See* bacterial wilt of brinjal, tobacco and tomato.

Pseudoparenchyma A fungal or algal tissue resembling parenchyma but made up of interwoven hyphae (fungi) or filaments (algae). The stripe of the mushroom and the thallus of red algae (*e.g.* Porphyra) are pseudoparenchymatous tissues.

Pseudopeziza *See* Heliotiales.

Pseudopeziza medicaginea *See* common leaf spot of alfalfa.

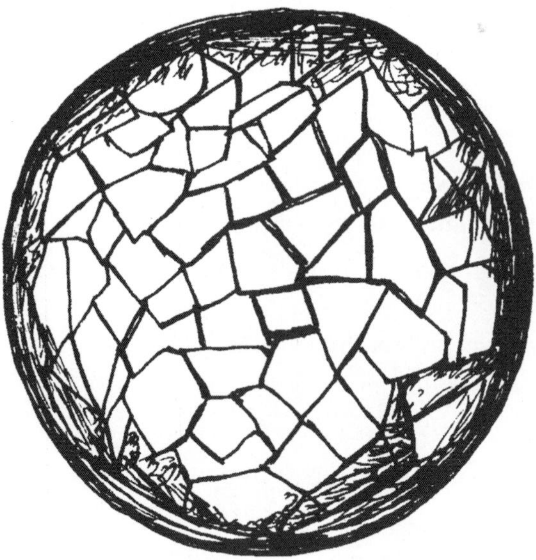

Pseudoparenchymatous Tissue
(Sclerotium of *Rizoctinia bactecicola* in section)

Pseudopodia *See* amoeboid movement.

Puccinia cagathami *See* rust of safflower.

Puccinia chrysanchemai *See* rust of Pyrethrum.

Puccinia graminis-avenae *See* stem rust of oats.

Puccinia glumarum Same as stripe rust of wheat and yellow rust of wheat and barley.

Puccinia graminis *See* berberry, *Berberis vulgaris.* Also *See* heterecious rusts.

Puccinia graminis tritici *See* stem rust of barley (*Hordeum vulgare*) and wheat (*Triticum aestivum*).

Puccinia helanthi In this organism, the haplont hyphae of the monosporidial origin are known to anastomose with the uredial hyphae that are diakaryotic.

Puccinia hordei *See* leaf rust of barley.

Puccinia kuehnii *See* rust of sugarcane.

Puccinia opera *See* rust of Job's tear millet.

Puccinia operta The new name give to *Uredo operta* on finding the telial stage in the *Uredo operta.*

Puccinia penniseti *See* rust of pearl millet.

Puccinia porri *See* rust of onion.

Puccinia purpurea It can attack jowar (Sorghum) only.

Puccinia recondata Same as leaf rust of wheat.

Puccinia sorghi Shows a host specificity by attacking maize only. It cannot attack sorghum.

Puccinia sorghii *See* leaf rust of maize.

Puccinia striiformis Same as yellow rust of wheat.

Puccinia striformis *See* stripe rust of barley and wheat.

Puccinia suaveolens It is a rust disease which does not have any aerialstage.

Puccinia substriata *See* rust of kodo millet.

Puccinia triticina It is the brown rust of wheat. Also known as leaf rust of wheat. Puckering Symptoms showing folds or wrinkles.

Puccinia graminis tritici

(a) Pycinidal and aecidial stage in section of barberry leaf; (b) Uredostage on wheat; (c) Uredospore germinating; (d) Part of a teleutosorus; (e): A teleutospore showing the germ pores; (f): A germinating teleutospore

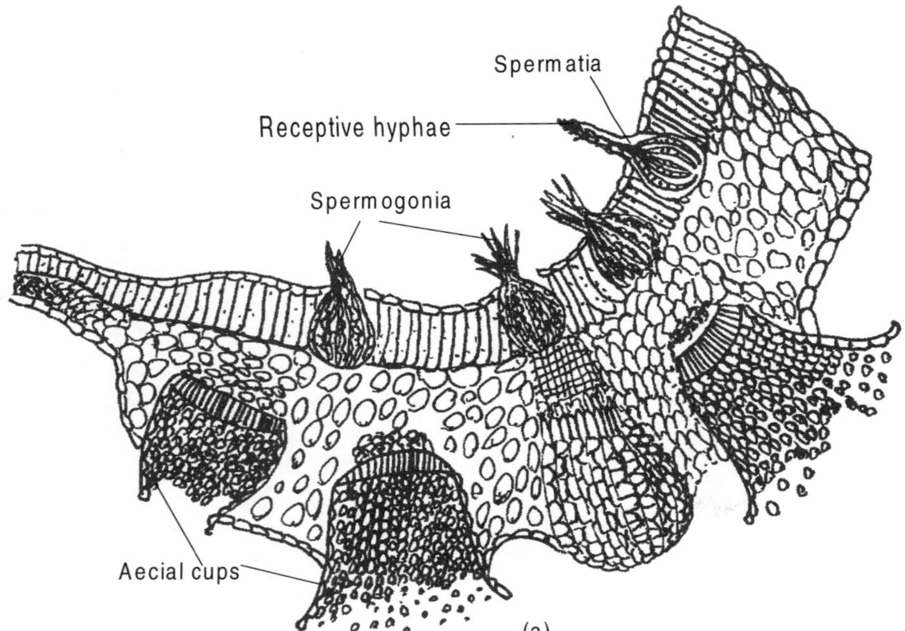

(a)

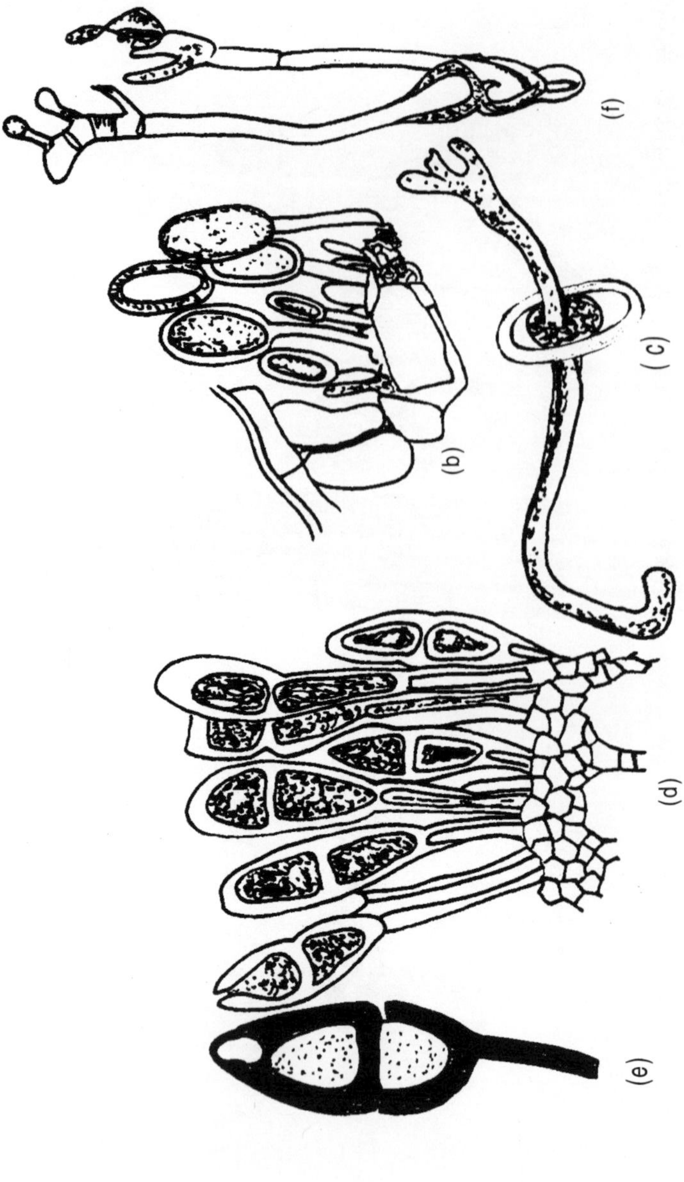
(b)
(c)
(d)
(e)
(f)

Puff-balls These beong to Eubasidiomycetes fungi.

Pullorum disease *See* Animal Science.

Pustules If the acervuli are formed below the surface of the host plant and limited in outline, they are termed as sori or pustules.

Pycnia Also called pycnidia. The spore bearing hyphae are sometimes enclosed in roundish cup or flask shaped receptacles known as Pycnia or Pycnidia and spores within them are known as Pycniospores or Pycnidiospores.

Pycniospores *See* Pycnia. Also called Pycnidiospores.

Pyliospore The morphology of teliospore furnishes the principal characteristics for the differentiation of genera of the Ureginales. Teliospores are the dormant spores and germinate as a rule after a period of rest by the formation of a long promycelium which is usually four-celled. Each one of the promycelial cells is formed. Each one of the promycelial cells is provided with a haploid nucleus formed as a result of the discussion of the fusion nucleus, the first being a reduction division of a short, pedicel like structure, the strigamata. Several deviations from this common method have been observed. Teliospores of Catenulospora germinate without a rest period; those of Melamspora are sessile. Teliospores of Coleosporium for an internal prymycelium *i.e.* the teleospore becomes four-celled, each cell being a promycelial cell.

Pyronema confluens A homothallic species, the cells developing into antheridia become club-shaped and their nuclei divide and re-divide until there are one hundred or more.

Pyrrophyta *Cryptomonads dinoflagellates* etc.

Pythiaceae A sub-class of Oomycetes. Sporangiospore little differentiated from vegetative hyphae; usually intracellular without haustoria.

Pythium aphanidermatum* and *Phytophthora palmivora *See* damping off of Tobacco.

Pythium aphanidermatum Causes damping off diseases in tobacco seedlings. Also *see* soft rot of ginger.

Pythium gramnicolum It causes a foot rot of wheat seedlings. Also *See* Pythium root rot of wheat.

Pythium root rot of wheat (*Triticum aestivum*) Caused by *Pythium gramnicolum*. Roots stunted and rotten; leaves pale green or brown.

***Pythium* spp.** *See* wilt of black pepper, clump rot of cardamom, damping off of brinjal and Pyrethrum.

Q

Quantity of fungi in soil Moulds are found in all horizons of the soil profile, the greatest number being in the A layers of the soil where organic matter is plentiful and aeration adequate.

R

Ramularia areola *See* grey mildew of cotton.

Ranikhet disease This disease of poultry is caused by Newcastle disease virus (NDA) belonging to the genus *Paramyxovirus* of the family *Paramyxoviridae.* This is the acute and fatal infection of chicken of all caused by the most virulent form of the virus. The Doyle's form (Asiatic form/velogenic viscerotropic RD) and Beach's (pneumoencephalitis) form tare the two types causing the diseases. They have separate symptoms for each infection. Proper hygiene and management are necessary to avoid introduction of the disease and to prevent the dissemination of the virus. In disease free areas RD is controlled by slaughter policy and immunoprophylaxis using vaccines. In areas where disease is endemic as in India, the most practical method to control is vaccination.

Reaction to stimuli The effect of acidity and alkalinity on the growth of fungi has been investigated. Fungi tolerate wide range of hydrogen and hydroxyl ion concentrations.

Red leaf of cotton (*Gossypium* spp.) It is a physiological disease. It occurs in American varieties, leaves turn yellow or red and roll downwards; when severely affected, whole leaf blade becomes red, leaving the veins green; later leaves dry up and shed.

Red leaf spot of Johnson grass Same as for Sorghum.

Red leaf spot of sorghum (*Sorghum vulgare*) Caused by *Colletotrichum graminicola.* Irregular tan coloured spot with reddish-brown margin; studded with black dot-like fructifications; appear on leaves; sometimes rotting of roots and stem is also noticed. For control, grow resistant varieties; practise crop rotation and improve soil fertility.

Red leaf spot of Sudan grass (*Sorghum sudanense*) Same as for sorghum.

Red rot of sugarcane (*Saccharum officinerum*) Caused by *Glomerella tucumanesis (Colletotrichum falcatum).* Leaves wither; cane shrinks showing black specks on shrivelled rind; split can gives alcoholic smell and shows red tissue with white cross bands. Grow resistant varieties; rogue out affected canes; use healthy seed.

Red stripe of sugarcane (*Saccharum officinerum*) Caused by *Cephaleuros parasiticus.* Small water-soaked green spots and yellow hairs appear on leaf and green twigs; twigs harden prematurely forming scaly patches which crack. Improve general cultivation and manuring; spray with Bordeaux mixture during monsoon or lime sulphur during cold weather before formation of hairs.

Reproduction in bacteria *See* Animal Science.

Resistance The inherent ability of plant to prevent or restrict the establishment and subsequent activities of a potential pathogen. It may be regarded as opposite to susceptibility.

Resupinate In the simplest sporophore, the fructification is a thick hyphal mat which grow radially in a definite manner, bears the hymenium on the lower side, and differentiates new hyphae and basidial elements at the peripher. This type of fructification is known as resupinate.

Rhinosporidiosis *See* Animal Science.

Rhizoctonia bataticola *See* root rot of cotton.

Rhizoctonia Its species are important among such fungi causing damping off diseases.

Rhizoctonia solani *See* sheath blight of paddy.

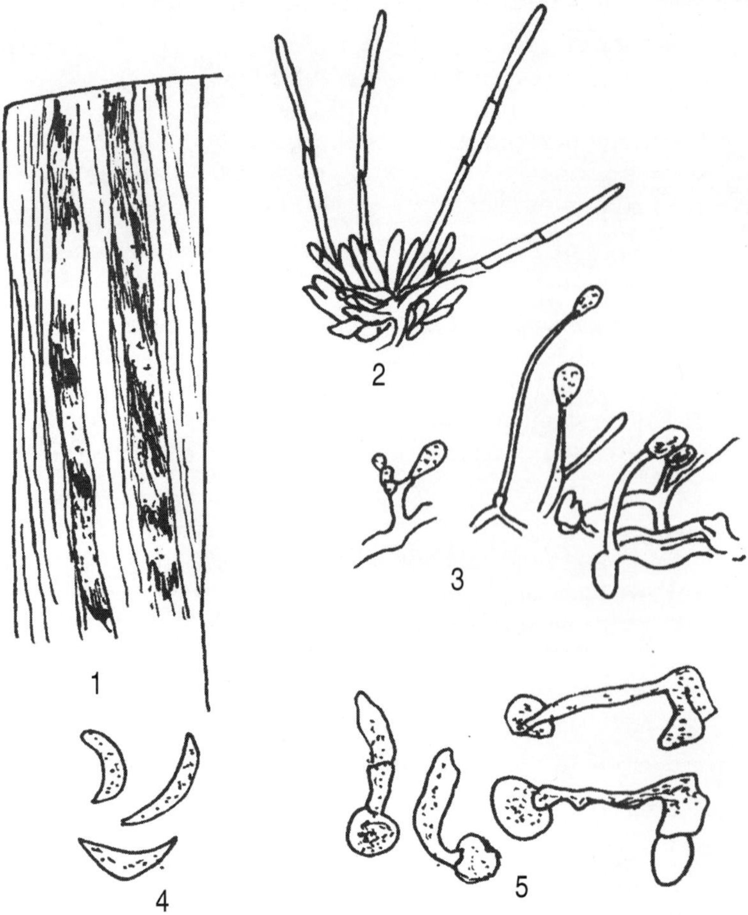

Red Rot of Surgercane *Colletotrichum falcatum*

1: Leaf showing lesions; 2: Acervulus showing setae, conidia and conidiophores; 3: Conidiophores and conidia; 4: Spores; 5: Germination of conidia with formiation of approsoria

Rhizoid One of the functions of the mycelium is that of absorbing nutrition. A part of it may penetrate the substratum as root like, filamentous and finally branched structure known as rhizoids. These are also known as false roots.

Rhizomorphs A specialised form of plectenchyma is the sclerotium. This is more or less rounded or cushioned shaped structure. It

Rhizoids of *Blastocladia pringsheimii*

consists of plexus of thick walled hyphae, the interior cells being hayline and isodiametric, the exterior dark-brown and crust-like complex are the rhizomorphs.

A structure found in certain fungi *e.g.* Agaricus, consisting of a mass of hyphae forming a complex strand of tissue that serves to transport food from one part of the mycelium to another, thus aiding propagation.

Rhizopus arrbizus This fungus causes fruit rot in apples.

Rhizopus artocarpi *See* inflorescence rot of Jack fruit.

Rhizopus nigricans *See* soft rot of sweet potato. It is common mould. In the older mycelium, three types of hyphae occur. Those of the first sort anchor the plant by means of rhizoids. Then are stolons and runners which produce another group of rhizoids.

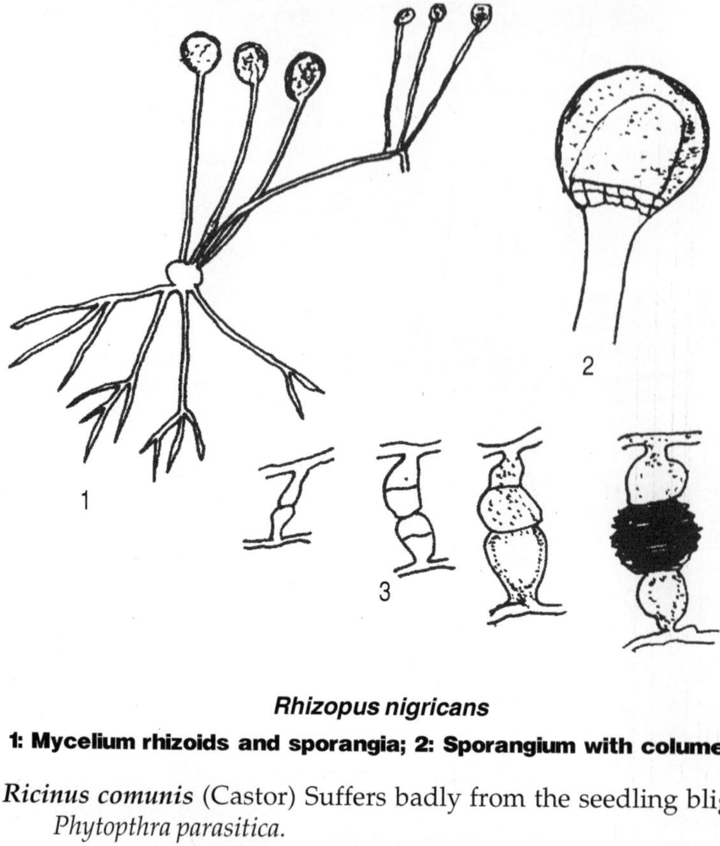

Rhizopus nigricans

1: Mycelium rhizoids and sporangia; 2: Sporangium with columella.

Ricinus comunis (Castor) Suffers badly from the seedling blight, *Phytopthra parasitica.*

Rind disease of sugarcane (*Saccharum officinarm*) caused by *Pleocyta sacchari.* Irregular dark, discoloured areas are formed on stem; pith rots and internodes shrink; black coiled threads appear on rind.

Rinderpest It is the most destructive of the viral diseases of cloven footed animals *viz.* cattle, buffaloes, sheep, goats, pigs and wild ruminants. The role of sheep and goats in the perpetuation of the virus and transmission of the disease has given increasing evidence to the presence of *Peste des petitis rumunants* (PPR) virus in outbreaks suspected to rinderpest in many parts of the country.

Ripe fruit rot Of banana caused by fungus *Gloesporium musarium;* can be controlled by spraying with Bordeaux mixture.

Ripe rot Of mango caused by a species of *Aspergillus.*

Root rot of tea (*Camellia sinensis*) Caused by *Botryodiplodia theobromae.* Bushes fail to produce new shoots and die off completely, or grow short shoots with mottled leaves; roots blacken and rot; black spherical fruit bodies are found embedded in bark. Treat soil with lime (2 lb per bush) during the early stage of the disease (mottling of new leaves); burn all infected material.

Root rot of coconut (*Cocos nucifera*) Caused by *Ganoderma lucidum.* Withering and drooping of leaves which remain suspended from tree; new leaves successively smaller and yellowish in colour; tree becomes barren. Fell diseased palms, dig out and burn; spread 2 to 3 lbs of sulphur and incorporate it into soil.

Root disease of coconut (*Cocus nucifera*) Outer whorl of leaves droops; leaflets become brown and dry; new leaves become progressively smaller; immature nuts shed. Better cultivation and manuring can control the disease.

Root rot and foot rot of barley (*Hordeum vulgare*) Caused by *Helinthosporium sativum.* Roots and collar show browning; later, elongated brown spots appear on leaves, glumes and spikeltes; ears poorly filled; grains shrivelled. Treat seed with Cerasan or Agrosan GN before sowing; grow resistant varieties; practise crop rotation and sanitation.

Root rot of arecanuts and beetle nuts (*Areca catechu*) Caused by *Ganoderma lucidum.* Same as for coconut.

Root rot of clover (*Trifolium* spp.) Caused by *Pellicularia filamentosa.* Same as for cowpea.

Root rot of cluster beans (*Cyamopsis tetragonoloba*) Caused by *Pellicularia filamentosa.* Same as for cotton.

Root rot of cotton (*Gossypium* spp.) Caused by *Rhizoctonia bataticola* and *R. solani.* Affected plants wilt suddenly; roots rot and show shedding park; disease in patches. Adjustment of date of sowing and growing of *Phaseolus aconitifolius* in mixture with cotton.

Root rot of cowpea (*Vigna sinensis*) Caused by *Pillicularia filamentos.* Same as for *Phaseolus* spp.

Root rot of gram (*Cicer arietinum*) Caused by *Opercullela padwickii.* Affected plants dry up from top downwards; leaves become pale green and later turn yellow and finally drop off; collar portion of plant becomes dark brown. Rogue out the affected plants.

Also caused by *Pellicularia filamentosa.* Same as for groundnut.

Root rot of grapes Caused by the fungus *Dematophora necatrix.*

Root rot of groundnut (*Arachis hypogea*) Caused by *Pellicularia filamentosa.* Red brown discolouration on stem which spreads and causes plant to die; tissues blackened and numerous tiny black sclerotia formed both inside and outside the tissue; sclerotia also formed on nuts underground. Practise crop rotation and good cultivation; treat seed with Agrosan GN before sowing.

Root rot of horsegram Same as for cowpea.

Root rot or charcoal rot of tomato (*Lycopersicum esculentum*) Caused by *Macrophomina phaseoli.* Same as dry root rot of brinjal.

Root rot of tea (*Camellia sinensis*) Caused by *Botryodiplodisa theobromae.* Bushes fail to produce new shoots and die of completely, or grow short shoots with mottled leaves; roots blacken and rot; black spherical fruit bodies are found embedded in bark. Treat soil with lime (2 lbs per bush) during the early stage of the disease (mottling of new leaves); burn all infected material.

Root-rot This disease is caused by the attack of fungi on the tap and secondary roots.

Rosellinia bunodes *See* black root rot of black pepper.

***Rosellinia* sp.** *See* black rot of coffee and tea, and black root rot of Jack fruit.

Rosette disease of groundnut (*Arachis hypogea*) Caused by virus. Affected plants show a dense clump of fusty dwarfed shoots with yellow leaves and erect habit.

Rotting The seeds of peas have been found rotten due to attack of Fusarium in the pre-emergence stage. Such rotting is distinct from the rotting of non-viable seed due to mould action. Seeds sown in moist soil may be rotten by the action of pathogenic fungi.

Roup in chicken *See* Animal Science.

Rust of alfalfa (*Medicago sativa*) Caused by *Uromyces striatus.* Small round or oblong dark brown rust pustules formed on leaves and petioles. Grow resistant varieties.

Rust of broad bean (*Vicia faba*) *Uromyces fabae* is the causal organism. Affected plants show yellow cup shaped erumpent pustules in clusters on all green parts of plant, mostly on the under surface of leaves; later, rust coloured small round pustules appear and then come up as jet black pustules.

Rust on castor (*Ricinus communis*) Caused by fungus *Melompsora ricini.* Rust pustules formed in small circles chiefly on lower surface of leaves; containing orange yellow powdery spore mass covered by epidermis; upper surface of leaf marked by corresponding small roundish-yellow spots. Grow resistant varieties.

Rust of clover (*Trifolium* spp.) Caused by *Uromyces trifolii* and *Uromyces minor.* Small brownish, rust pustules appear on all green parts. Grow resistant varieties.

Rust of cotton (*Gossypium* spp.) Caused by *Ceratilium desmium.* Innumerable, small, yellowish-brown rust pustules formed, on leaves. The disease is seldom serious.

Rust of cowpea (*Vigna sinensis*) and horsegram Caused by *Uromyces appendiculatus.* Same as for *Phaseolous* spp.

Rust fenugreek (*Trigonella foemum-graecum*) Caused by *Uromyces anthylidis.* Same as for Alfalfa.

Rust of figs (*Ficus carica*) Caused by *Cerotelium fici.* Small round raised, brown to black, lesions on leaves often causing leaf fall. Dust under surface of foliage with sulphur or spray 1 per cent Bordeaux mixture.

Rust of gram (*Cicer arietinum*) Caused by *Uromyces cicerisarietini.* Rust pustules are seen chiefly on lower surface of leaf as small round oval, brown coloured bodies, which in later stages become dark brown; affected leaves often dry. Grow resistant varieties.

Rust of grape (*Vitis vinifera*) Caused by *Phakopspora vitis.* Dark numerous orange coloured sori appear on lower surface of leaves of variety Black prince. Dust with finely powdered sulphur.

Rust of Italian millet (*Setaria italica*) Caused by *Uromyces setariaeitalicae.* There are numerous rust pustules, fall of powdery spot mass; formed on both surfaces of leaves; colour of pustules brown in the beginning, but turns black at later stage. For control, grow resistant varieties.

Rust of job's tear millet (*Coix lachryma-jobi*) Caused by *Puccinia operala.* Small straw-coloured rust pustules found scattered on surfaces of leaves. Grow resistant varieties.

Rust of kodo millet (*Paspalum scrobiculatum*) This disease is caused by *Puccinia substriata and Uredo paspaliscrobiculati.* Pustules cinnamon brown to chocolate brown in colour; appear scattered on both surfaces of leaves. In case of *U. paspali-scrobiculati,* small, scattered brown rust pustules appear mostly on upper surface of leaves.

Rust of lablab bean (*Dolichos lablab*) Caused by *Uromyces appendiculatus.* Several minute, slightly raised spots appear on leaves; spots develops into brownish powdery sori which become dark brown or black. Dust crop with finely powdered sulphur @ 15 lbs/acre.

Rust of lathyrus (*Lathyrus sativus*) Same as of broad beans.

Rust or leaf disease of coffee (*Coffee arabica*) Caused by *Hemileia vastatri.* Orange yellow spots with spores appear on surface of leaves shed. Grow resistant strains; spray under surface with Bordeaux mixture (2–2–50), one month after occurrence of blossom showers, and later during break becomes in S.W. and N.E. monsoons.

Rust of lentil (*Lens esculenta*) Caused by *Uromyces fabae. See* in broad beans.

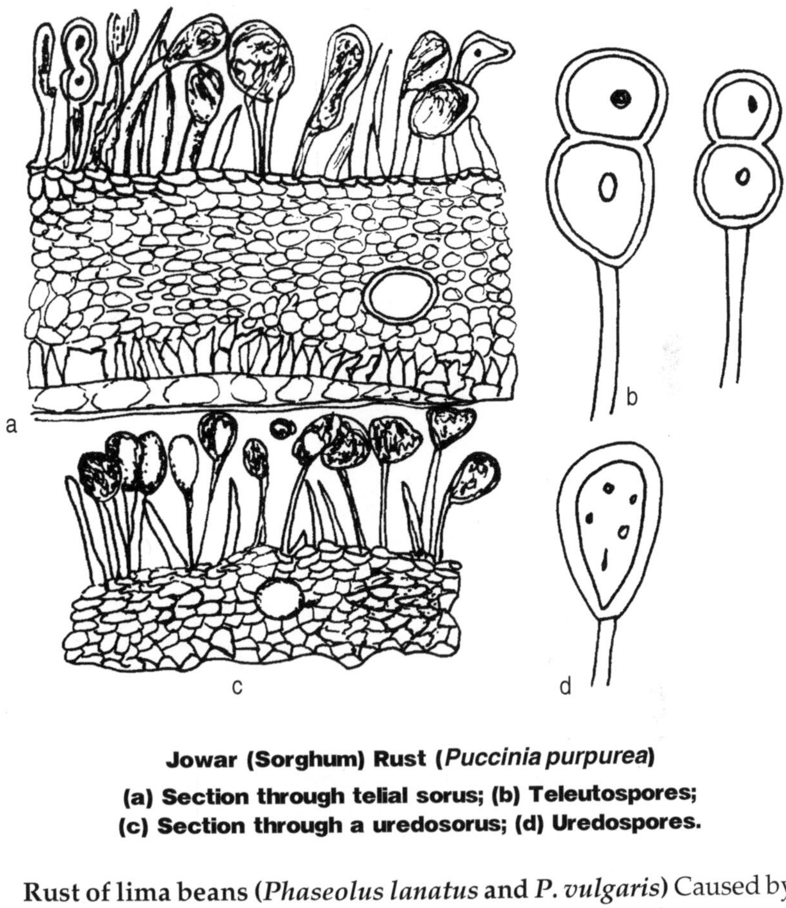

Jowar (Sorghum) Rust (*Puccinia purpurea*)

(a) Section through telial sorus; (b) Teleutospores; (c) Section through a uredosorus; (d) Uredospores.

Rust of lima beans (*Phaseolus lanatus* and *P. vulgaris*) Caused by *Uromyces appendculatus.* Control same as in *Phaseolus* spp.

Rust of linseed (*Linum usitatissimum*) Caused by *Melampsora lini.*

Rust of maize caused by *Puccinia maydii.*

Rust of onion (*Allium cepa*) Caused by *Puccinia porri.* Longitudinal lesions which, on rupturing, expose yellowish uredospores; telio are produced later intermingled among uredia bearing black teliospores.

Rust on Johnson grass Same as for sorghum.

Rust on velvet beans (*Mucuna* sp.) Caused by *Uromyces mucunae.* Powdery, rust pustules, brown to black in colour found scattered in large number on under surface of leaves. Control measures are same as for Phaseolus.

Rust of pearl millet (*Penniseum typhoideum*) Small, brownish-yellow rust pustules formed in groups of leaves and stems, which become black later. For control, grow resistant varieties. The causal organism is *Puccinia penniseti.*

Rust of peas (*Pisum sativum*) Caused by *Uromyces fabae* and *U. pisi.* Uredia and telia produced on host. Dust crop with sulphur @ 15–20 lbs/acre.

Rust of phaseolus sp. (*Phaseolus radiatus* and *Phaseolus mungo* etc.) Caused by *Uromyces appendiculatus.* Numerous brown erampent pustules, mostly on the under surface of leaves; sometimes stems and pods also affected. For control, grow resistant varieties.

Rust of pyrethrum (*Chrysanthemum cinarariaefolium*) Caused by *Puccinia carthami.* All green parts of plant attacked but infection more on leaves; scattered, small, round, rust pustules with ruptured epidermis formed on both surfaces of leaves. For control, treat seeds with Agrosan GN @ 2.5 gms/kg of seed; destruction of (Pohli) *Carthmus oxycantha* seeds.

Rust of sann-hemp (*Crotoloria juncea*) Causal organism is *Uromyces decoratus.* Rust pustules appear on all aerial parts as yellowish brown scattered pustules, later becoming dark brown; plant tissues dry up.

Rust of sorghum (*Sorghum vulgare*) Caused by *Puccinia purpurea.* Bright purplish red spots appear on leaf surface; rust pustules formed in these spots especially on lower surface. For control, grow resistant varieties.

Rust of sugarcane (*Saccharum officinarum*) Caused by *Puccinia kuehnii.* Minute elongated orange pustules, later turning brown appear on under surface of leaves. Grow resistant varieties.

Rough spored bunt This is bunt of wheat. The causal organism is *Tilletia caries.*

S

Saccharomyces (Yeasts) A genus of unicellular ascomycete fungi that can live in both aerobic and anaerobic conditions. They are important in the brewing and baking industries respectively for the alcohol and carbon dioxide they produce by anaerobic respiration. Reproduction is generally asexual by budding although in adverse conditions sexual spores maybe formed.

Saccharomycetaceae In this family of Endomycetales, however, the mycelium is absent.

Saccharomyces cerevisiae This is the best known species among yeasts. It is common brewer's and baker's yeast.

Saccharomyces ellipsoideus This is found on grapes in nature and is extensively used in wine-making.

Sann-hemp wilt The temperature for the maximum growth of the fungus is 25–28°C. The causal organism is the *Fusarium crotalarae.*

Saprolinales One of the orders of the sub-class Oomycetes of Phycomycetes.

Saptoria tritici *See* leaf blotch of wheat.

Sapwood (Alburnum) The outer living xylem cells in a tree trunk, consisting of xylem elements, parenchyma, and medullary rays, that are actively involved in water transport and food storage. Compare heartwood.

Scab of apples It was first named as *Fusicladium dendriticum* and later it was found that it was the imperfect state of *Venturia inaequalis.*

Scab Of Citrus. Often mistaken for canker. The lesions appear on the same portions and are somewhat similar in appearance. Scab lesions project on only one side of the leaf, where as those of canker are raised on both sides and scab lacks the characteristic yellow halo. On both leaves and fruits, the lesions are generally accompanied by some distortions.

Scab of potato (*Solanum tuberosum*) Caused by *Streptomyces scabies.* Superficial roughened areas of corky tissue, often slightly below the plane of healthy skin of tubers. Control as for black scurf.

Scalecosporae A family of fungi *imperfecti,* spores acicular, to filiform, or worm shaped, one to many celled.

Schizomycetes The division of Thallophyta has one of the classes as this and includes bacteria.

Schizosaccharomyces It is a genus of yeasts. *See* saccharomyces.

Sclerophthora macrospara *See* downy mildew of finger millet.

Sclerospora graminicola *See* downy mildew of common millet and pearl millet's green ear disease.

Sclerospora philippinenisis Oospores have not yet been observed in this species. *See* downy mildew of maize.

Sclerospora sorghi *See* downy mildew of sorghum.

Sclerotial disease of barley (*H. vulgare*) Caused by *Pillicularia rolfsii.* White weft of mycelium formed near the base of the plant on which white to reddish-tan spherical sclerotia appear. For control, soil sterilization and rotation practices needed.

Sclerotial disease of cotton (*Gossypium*) Caused by *Pillicularia rolfsii. See* in barley.

Sclerotial disease of cluster beams *Pillicularia rolfsii.* Same as for cotton.

Sclerotial disease of jute (*Torchorus* spp.) Caused by *Pellicularia rolfsii.* Same as for barley.

Sclerotium As specialised for of plectenchyma is the sclerotium. This is one or less rounded shape or cushion-shaped structure. The resting body of certain fungi *e.g.* ergot, formed from a mass of hyphae. *See* also stroma.

Sclerotinia sclerotiorum Stem rot disease of rocket salad and Rosalle. Also see gram and rocket salad or jambo.

Sclerotinia *See* Helotiales.

Sclerotial root rot of finger millet (*Eleucine coracana*) Caused by *Pillicularia rolfsii.* Affected plants dwarf and pale; dense white, fan-shaped weft of mycelium appears on clump near soil; later brown sclerotia formed in threads of mycelium. Practise clean cultivation for control of disease. Also a disease of barley.

Scrapie *See* Animal Science.

Seath blight of paddy (*Oryza sativa*) Causal organism is *Rhizoctonia solani.* Lesions white, surrounded by irregular dark brown margin of leaf-sheath at the water level; leaf blade of affected sheath dries up from tip downwards.

Secondary spores To increase the number of spores, and therefore the chances of survival of the fungus, several devices have developed among the fungi and one of them is production of secondary, tertiary spore.

Seed borne infection In loose smut of wheat and barley, the infection of the spores is externally seed borne, external application of seed disinfectants is of no use whatsoever.

Seedling blight and damping off of cinchona (*Cinchona* sp.) Caused by *Phytophthora* sp. and *Macrophomina araceae.* Damping off of seedlings in nursery encouraged by over-crowding and too much shade. Remove excess shade; restrict watering; spray seedlings with Perenox (2 ozs. in two galls or Yellow Cuprocide 1.5 ozs. in two galls) per 24 sq. ft. of nursery.

Seedling blight of castor (*Ricinus communis*) Caused by *Phytophthora parasitica.* Dull green, round patches appear of both surfaces of leaf which further spread; spots turn yellow or brown with concentric zones of brown colour on lower surface of leaves and also on affected stems. Do not grow the crop on low lying areas or damp places.

Seedling blight of sesbania (*Colletotrichum capsici*) Causes the disease. Collar region of the affected seedlings show elongated or oblong cankers with black bristle like tufts of setae. For control, spray with Bordeaux mixture.

Seedling disease In the case of seed-borne diseases, the disease organism or its spores are carried either externally on the surface or internally within the seed. When it is external, it may be destroyed by treating the seed with a chemical *e.g.* formalin, sulphur, copper carbonate, organo-mercurials like Agrosan, Ceresan, etc. Except formalin, all other chemicals are used in dust form. The chemical thoroughly mixed with the seed, preferably in a rotating drums so as to cover the seed surface with.

Seedling disease of pea (*Pisum sativum*) Caused by *Pythium* spp. Pre-emergence and post-emergence damping off; fibrous roots infected. Treat seed with Cerasan, Arasan or Spergon at 20 ozs. per bushel 40-50 lbs.

Septa *See* mycelium.

Septoria chrysanthemlella *See* leaf blotch of *Pyrethrum.*

Septate divided or partitioned hyphae. Hyphae are terminal or apical elongation. They are segmented into cells by cross-walls which are called septa and the segmented mycelium is known as septate.

Septoria A member of Sphaeropsidales causing some destructive plant diseases.

Septoria nodorum *See* glume blotch of wheat.

Short cycle rust *See* microcyclic rusts.

Shot-holes In some leaf spots, the dead tissue of the spot is sometimes shed having a circular perforation called a shot-hole.

Shoot smut of barnyard millet (*Echinochola frumentacea*) Caused by *Ustilago crussgalli.* Large smutted galls containing black mass of spores occur in inflorescence, and on stem especially on nodes; also found on young shoots and in axils of older leaves; shoot infection causes deformity resulting in twisted mass of leafy shoots with sometimes aborted ears.

Septate Mycelium

Shoot rot of banana, plantain (*Musa* spp.) Caused by *Ceratostomella paradoxa.* Brown or dark water soaked patches on the bulbs.

Silk rot of maize Caused by *Pythium planidermatum.*

Smooth spored bunt The causal organism is *Telletia foetida.* Smuts basidiomycete fungi of the order Oscilaginales. Many are important parasites of cereals and form a mass of sooty black spores in place of the grain. Examples of smuts are Tilletia and Ustilago.

Smoking In this process, the food stuffs like pork and fish are subjected to the action of smoke for their preservation.

Smuts' infection They fall into three groups as regard the manner of infecting the host (1) Chlamydospores externally associated with the seed and infecting seedlings (2) Chlamydospores not externally associated with seed; infection taking place at the time of flowering; chlamydospores germinating on the stigma

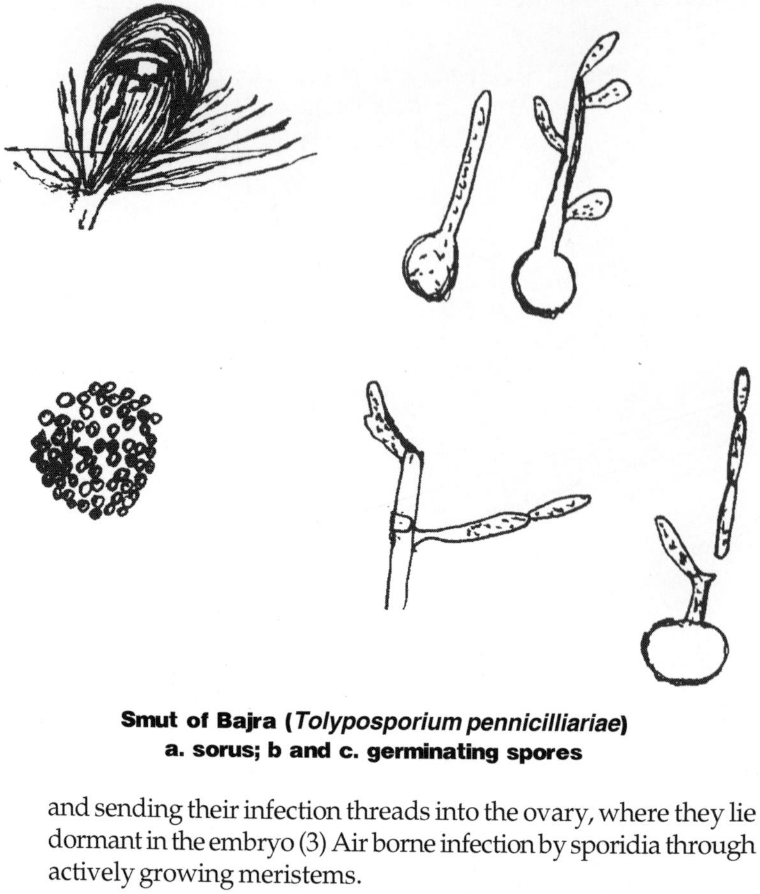

Smut of Bajra (*Tolyposporium pennicilliariae*)
a. sorus; b and c. germinating spores

and sending their infection threads into the ovary, where they lie dormant in the embryo (3) Air borne infection by sporidia through actively growing meristems.

Smut of Italian millet (*Setaria italica*) Caused by *Ustilago crameri.* Affected grains convert into black, loose spore mass which remains enclosed in a membrane till plant matures. Treat seed with Agrosan GN before sowing @ 2 gms/kg.

Smut of barnyard millet (*Echinochola frumentacea*) Caused by *Ustilago panicifrumentacei* and *U. paradox.* Grains in ears converted into black spores mass; infected grains, in case of *U. panicifrumentacei,* assume bigger size than normal grains; in case of *U. paradox* remains unchanged.

Smut of finger millet (*Eleusine coracana*) Caused by *Melanopsichium eleusinis.* Affected grains are deformed, and filled with compact, and dark coloured gall-like spore mass; found scattered in the spike.

Smut of kodo millet (*Paspalum scrobiculatum*) Caused by *Sorosporium paspali* var. *venucosum*. Affected ears wholly transformed into a black smut sorus; enclosed by a certain coloured membrane; in earlier stages; remains partly covered with flag leaf.

Smut of maize (*Zea mays*) Caused by *Ustilago zeae*. Soft tumours on all parts of the plant, mostly on cob; tumours when mature rupture and release black dusty mass of spores. Use resistant varieties, practise crop rotation and sanitation.

Smut of onion (*Allium cepa*) Caused by *Urocystis cepulae*. Dark slightly thickened area on cotyledons of young seedlings; numerous raised blisters near base of scales of older plants which, on rupturing, expose black powdery mass of spores. Place a mixture of lime and sulphur in furrow with the seed, apply as solution of formaldehyde (one pint of 40 per cent formaldehyde in 16 galls of water, applied at the rate of 1 gall to 150 ft. of furrow); for nursery, dress seed with 1 lb. of Arasan to 10 lbs. of seed.

Smut of pearl millet (*Pennisetum thphoideum*) Caused by *Tolyposporium pencillariae* and *T. senegalens*. Affected ears found scattered in ear, and project prominently beyond glumes.

Smut of job's tear millet (*Coix lachryma-jobi*) Caused by *Ustilago coicis* and *U. lachrymijobi*. By *U. coicis*, all the ovaries are destroyed, black spore mass held by hard floral glumes. In case of *U. lachrymaejobi*, only a few ovaries attacked and covered by glumes; much flower not attacked.

Smut of sugarcane (*Saccharum officinarum*) Caused by *Ustilago scitaminea*. Growing shoot turns into long whip-like black growth, covered by powdery black spores enclosed in thin membrane. Remove diseased clumps and use healthy seed.

Soft rot of ginger (*Zingiber officinalis*) Caused by *Pythium aphanidermatum*. Rotting of rhizomes in field during rainy weather, which continues during storage. Plant healthy seed rhizomes; rogue out diseased plants; steep seed rhizomes in mercuric chloride (1 in 1,000) for 1.5 hours before storage.

Soft rot of sweet potato (*Ipomea batatas*) Caused by *Rhizopus nigricans*. Storage rot of tubers; tissue becomes soft and stringy.

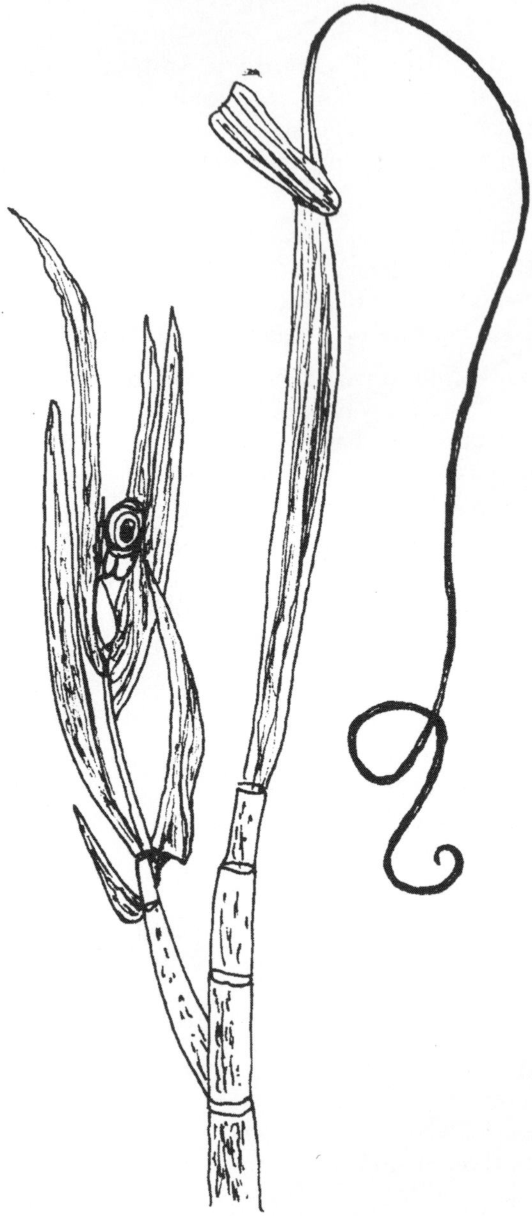

Smut Whip of Sugarcane

Avoid brushing tubers at harvesting; walls and floors of store room should be disinfected before crop is stored.

Soil-borne disease In the case of soil-borne diseases where the fungus living in the soil attacks the seeds or seedling, the use of soil disinfectants or sterilisers *e.g.* formaldehyde, formasan, etc. are helpful in destroying the fungus. Proving the parasitism of fungi causing wilts or foot-rots is a very laborious process. As these diseases are soil-borne, the soil in which host plants are to be raised has to be disinfected against the suspected pathogen.

Soil fungi It is well known that they play a very important part in the transformations of the soil constituents. A large number of species have been identified. Like the bacteria and actionomyces, fungi contain to chlorophyll and must depend for their energy and carbon on the organic matter of soil. The superficial characteristic that distinguish fungi is the filamentous nature of their vegetative forms.

The Fungal Mycelim

Sooty blotch and fly-speck disease of apples. It is caused by fungus *Leptothyrium pomi.* This is a fruit rot disease. Can effectively be controlled by lime-sulphur or colloidal sulphur sprays.

Sooty blotch of clover (*Trifolium* spp.) Caused by *Cymadothia trifolii.* Numerous dark brown or black angular blotches appear chiefly on lower surface of leaves; later blotches turn black and warty; affected leaves look yellow or brown.

Sooty mould Of mango, caused by *Capnodium mangiferum.*

Sooty mould of guava (*Psidium guava*) Caused by *Meliola psidi;* same as for Sapodilla.

Sooty mould of mango (*Mangifera indica*) Caused by *Capnodium ramosum.* Same as in Sapodilla.

Sooty mould of sapodilla (*Achrus sapota*) Caused by *Capnodium* sp. Sooty unsightly growth on upper surface of leaves. Spray with oil rosin soap (1 lb. in six galls of water) or crude oil emulsion (1 lb. in six galls) followed by starch solution (2 lbs. starch boiled in eight gallons of water); or organophosphorus compounds like Folidol, Ekaton or HEPT (4 ozs. in 2 galls of water).

Sooty stripe of sorghum (*Sorghum vulgare*) Caused by *Ramulspora sorghii.* Elongated lesions, regular in outline and tan-coloured with red margins appear on leaves; later, lesions become sooty coloured as black loosely attached sclerotia are formed. Rogue out the affected plants.

Sooty stripe on Sudan grass (*Sorghum sudanense*) Same as for Sorghum.

Sorghum downy mildew of maize Caused by *Peronosclerospora sorghii.*

Sori (*sing. sorus*) *See* pustules.

Sorosporium paslali *See* smut of kodo millet.

Spergon A non-metallic fungicidal dust chemically known as tetrachloro-p-benzoquinone. It is used as a seed dresser.

Sphacelotheca cruenta *See* loose smut of sorghum. Spores are rough.

Sphacelotheca reiliana Soil borne smut of jowar (sorghum). Also *See* head smut of sorghum and *Zea mays.*

Sphacelotheca sorghi *See* grain smut of sorghum. Spores are rough disease of sorghum.

Sphacelia sorghi Causes sugary disease in sorghum.

Spharotheca humuli *See* powdery mildew of cucurbits.

Sphaeropsidales Also known as Phomales or Phyllostictales, have a septate mycelium which is profusely branched. All members of this order, so far as present known ones belong to the various families of Ascomycetes.

Sphaeropsis malorum *See* black rot of apple.

Sphaerotheca pannosa *See* mildew of peach and apricot.

Spirillum In such bacteria, cells are in the form of helix or corkscrews. Very short spirilla with less than one full turn are called Vibrios.

Spirochaetosis *See* Animal Science.

Spirogyra It is a fresh water alga and is cosmopolitan in distribution.

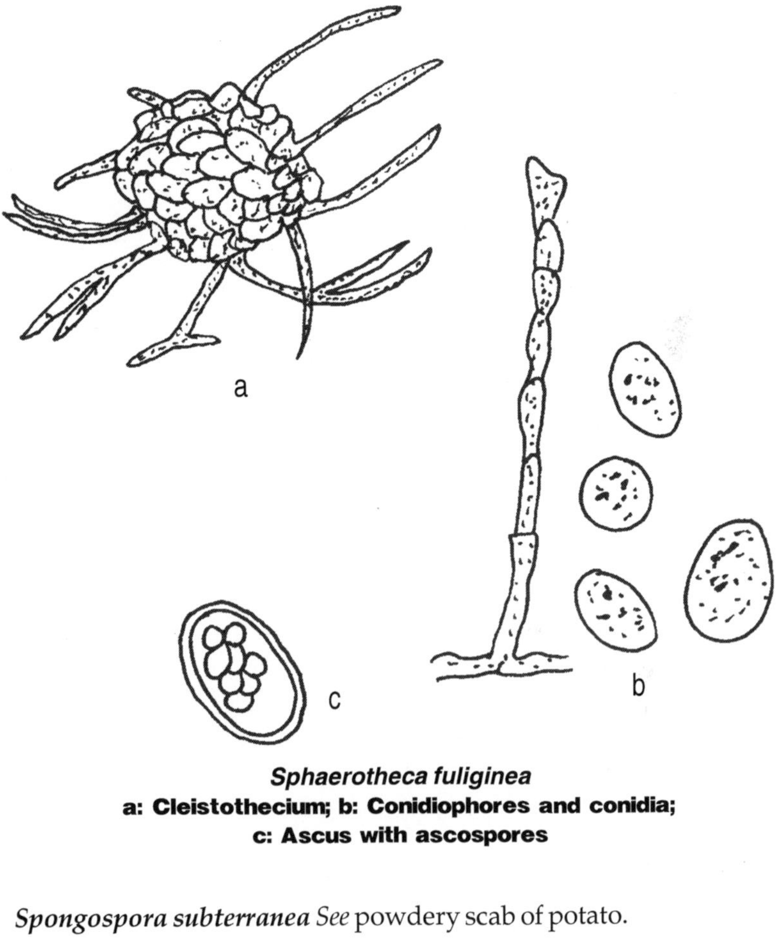

Sphaerotheca fuliginea
a: Cleistothecium; b: Conidiophores and conidia; c: Ascus with ascospores

Spongospora subterranea *See* powdery scab of potato.

Spongy tissue It consists of a few loosely packed cells with large intercellular spaces or cavities.

Spontaneous generation Over 2000 years ago, Aristotle proposed a hypothesis of spontaneous generation according to which the non-living matter can cause itself to become alive and this is due to the presence of some active principle or the capacity of doing something present in the bits of matter.

Spore-balls In *Urocystis brassicae,* induces gall formation in the roots of mustard. Chlamydospores are spherical to avoid and one-celled, but sometimes several chlamydospores may become united into more or less permanent spore-balls.

Spore-formers None of the bacteria that cause diseases in plants are spore-formers.

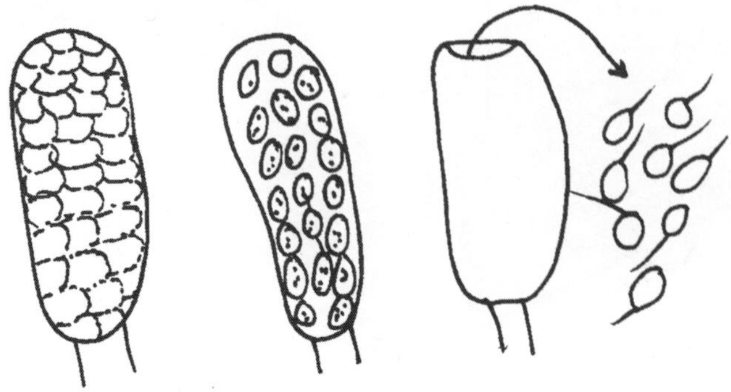

Sporangia and Ejection Monoflagellate Zoospores

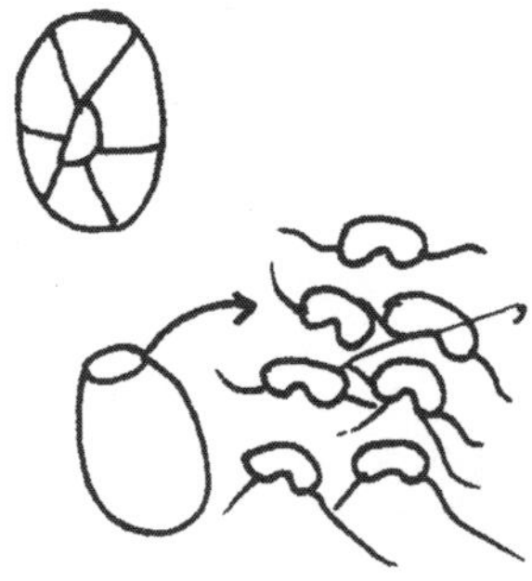

Sporangia Showing Cleavage and Ejection of Biflagellate Zoospores

Spore sac *See* Theca.

Spore-initials In *Rhizopus nigricans,* the spore pieces of protoplasm produced by the process of progressive cleavage, are the spore-initials.

Spore sac *See* Theca.

Sporidia The basiospores which are oval, elliptical or spherical, smooth or ehinulate, hayline or coloured; are also known as sporidia.

Sporotichosis A disease of horse, mule and donkeys. *See* Animal Science.

Sprays The use of sprays and dusts to protect plants against diseases is a standard method and practice.

Spray equipment and its care Spraying is the most common method of applying insecticides, fungicides and weedicides. Sprayers are classified as low-volume and high-volume sprayers. Low volume sprayers usually apply less than 30 gallons of total spray per acre. With aircraft, it is possible to apply as little as one gallon of spray per acre. Our small agricultural holdings coupled with the practice of growing mixed crops present difficulties to spraying of herbicides by aircraft. However, the foot-operated sprayers which are capable of delivering about 80–100 gallons of spray per acre are useful in that the drift hazard to sensitive crops is eliminated.

Spread of plant diseases *See* dissemination of plant diseases.

Stack burn disease of paddy (*Oryza sativa*) Caused by *Trichoconis padwickii.* Diseased seedlings show small, dark brown or black spots on roots; spots on leaves pale brown in colour with dark brown margins; those on glumes pink to reddish-brown in colour surrounded with a dark ring.

Stachylidium theobromae *See* cigar end disease of banana.

Stalk rot of maize Caused by *Earwinia chrysenthum.*

Staphylococcus Bacterial cluster of cells. These look like a bunch of grapes and are formed by repeated divisions of parent cell in all planes in an irregular order.

Staurosporae A family of *fungi imperfecti* in which spores stellate, one to many celled.

Stem and fruit canker of tomato (*lycopersicum esculentum*) Caused by *Corynebacterium michiganense.* Light brown streaks appear on leaf and leaf petiole. Seed treatment with hot water or organomercurials crop rotation.

Stem black A disease of apples, caused by *Conothecium chomatosphorium.* Lime sulphur spray can control the disease. This is also known as blister disease. *See* also blister disease of apple.

Stem bleeding, stem crack and sun scorch of arecanuts (*Areca catechu*) Caused by *Ceretostomella paradoxa*. Cover leaf sheaths on south-west side of young plants. *See* under coconut.

Stem bleeding of coconut (*Cocos nucifera*) Caused by *Ceretostomella paradoxa*. Reddish brown liquid oozes out of cracks on stem; fluid dried and turns black. Cut out discoloured tissue; paint cut surface with hot coal tar or Bordeaux mixture.

Stem brown A disease of apples caused by *Botryosphoeria brown ribis*. To control, spray twice with lime-sulphur.

Stem canker of cinchona Caused by *Botryodiplodia theobromae*. Cankerous growth in open wounds. Scrape out infected portion; apply Bordeaux mixture.

Stem canker of sugarcane (*Saccharum officinarum*) Caused by *Cytospora sacchari*. Pith of affected canes becomes hollow; shows black incrustation; mostly saprophytic. Use disease-free healthy seed; remove affected stubbles and cane debris.

Stem end rot of banana Caused by *Gloesporium* spp. and *Botrydiplodia* spp. Spray of Bordeaux mixture can control the disease.

Stem-galls of coriander (*Coriandrum satvium*) Caused by the organism *Protomyces macrospores*. It appears in the form of tumour-like swellings on the veins, leaf-stalks, peduncles and other green parts of the plant.

Stem rot of cucurbits Caused by *Pellicularia (Sclertium) rolfsii*. Cortical decay of stem at soil level, with production of mustard like sclerotia; young leaves wilt; watery rot of fruits, with abundant sclerotia. Practise crop rotation.

Stem rot or dry rot of jute (*Corchorus* spp.) Caused by *Macrophomina phaseoli*. Roots become brown and discoloured below bark; brown spots formed at collar region on which black-dot like structures, pycnidia, appear; latter collar rots, plant sheds its leaves and dies. To control, resistant varieties, practise crop sanitation; use clean seed. Treat seed with Agrosan GN or Cerasan @ 2–2.5 gms/kg seed.

Stem rot of gram (*Cicer arietinum*) Caused by *Sclerotinia sclerotiorum*. *See* rocket salad or jambo stem rot.

Stem rot of paddy (*Oryza sativa*) Caused by *Leptosphaepria salvinii.* Usually base of clumps affected; stem breaks; grains partially filled; numerous black sclerotia found all over greyish weft of mycelium within the hollow stem in basal node.

Stem rot of papaya The causal organism is reported as several species of *Pythium* but *Pythim aphanidermatum* is said to be most responsible for this disease. The disease can be avoided by planting papayas on a well drained soil. Also called collar rot or foot rot, caused by *Pythium aphenidermatum* and *P. ultimum.*

Stem rot of rocket salad (*Euraca sativa*) Caused by *Sclerotinia sclerotiorum.* Scattered, elongated, concentric zonatic lesions on stem; affected portions covered by white, cotton mycelium with black sclerotia.

Stem rot of rozelle (*Hibiscus subdariffa*) Large brown patches on stems and branches; large black elongated sclerotia formed in the pitch region at point of infection. For control, practise crop sanitation; give deep ploughing; spray crop with 1 per cent Bordeaux mixture. The disease is caused by the fungus *Sclerotinia sclerotiorum.*

Stem rot of tobacco (*Nicotiana* spp.) Caused by *Pellicularia (Sclerotium) rolfsii.* Girdling of stem at ground level; infected plants become yellow and frequently break over; mustard-like sclerotia produced. Remove and destroy infected plants; prevent contact of irrigation water with stem.

Stem rust of barley (*Hordeum vulgare*) and wheat (*Triticum aestivum*) Caused by *Puccinia graminis tritici.* Reddish brown to dark brown oblong pustules on clumps and leaf sheath; later pustules turn black; grains shrivelled. For control, grow resistant varieties; dust with sulphur @ 15–20 lbs./acre or spray with parate or zinc sulphate @ ½ gallon plus ¾ lbs./100 gallons of water/acre.

Stem rust of oats (*Avena sativa*) Caused by *Puccinia graminis-avenae.* Similar to stem rust of barley; *Vulpia myuros* and *Briza minor* common wild grasses act as alternate hosts. For control, dust with sulphur; grow resistant varieties.

Stem rust of wheat (*Triticum aestivum*) Caused by *Puccinia graminis tritici.* Same as of barley.

Stenosis of cotton (*Gossypium* spp.) Caused by a virus. Leaves of affected plants get crinkled and deformed showing yellow or pink discolouration. Grow resistant varieties for its control.

Sterigma A finger-like projection upon which spores are formed in many fungi. In basidiomycetes fungi, four sterigmata, each bearing one basidiospore, are usually borne on each basidium. In certain ascomycetes fungi, *e.g.* Penicillium, the sterigmata give rise to chains of conidiophores.

Sterigmata In Basidiomycetes, sometimes after its formation, the synkaryon divides by a reduction division. This is followed by a second division, which is mitotic, four nuclei being thus formed, two of each sex. Later slender projections, known as sterigmata, develop at the distal end of the basidium on which basidiospores are produced, each with a single haploid nucleus. In *Aspergillus niger*, the vegetative cell from which a conidiophore is formed is known as foot cell. The tip of the conidiophore becomes swollen to form a club shaped or rounded structure called vesicle. Form the surface of the vesicle arise a large number of buds called sterigmata or phialides.

Sterile hyphae The fungul hyphae known as paraphyses.

Sterilization Although several species of bacteria can live in hot springs, most of them are killed in boiling water. The spores are not even affected by boiling water. The spores are killed by either the use of dry heat or steam under high pressure. Such a process which kills spores along with bacterial cells is termed as sterilization. Where a pathogen is a facultative parasite, it has to be brought to pure culture before it can be further studied. For this purpose, small pieces of diseased tissue a microscopic examination of which has shown the presence of the fungus are washed and surface sterilized by using silver-nitrate-sodium-chloride solutions.

Streptococcus Filaments of the cocci are formed by the repeated divisions of the parent cell in the same plane.

Storage rot of banana Caused by the spp. of *Gloesporium* and *Diploidia.*

Storage rot of pineapple (*Ananas sativus*) Caused by *Ceratostomella paradoxa.* A soft watery growth on upper surface of leaves. Smear cut-ends with antiseptic.

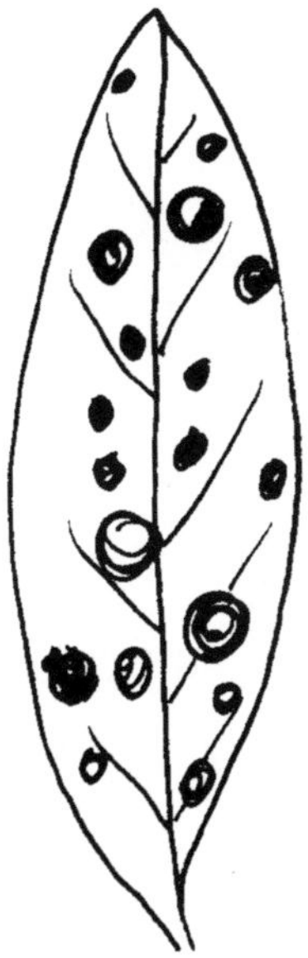

Shot Holes

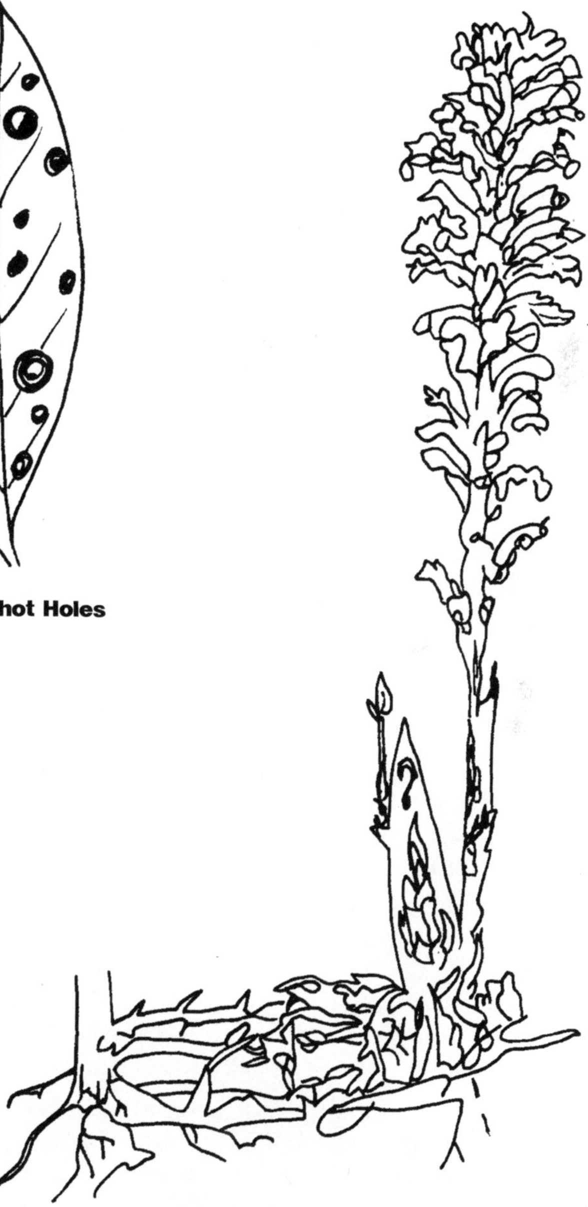

***Striga* sp. on Roots of Another Plant**

Stripe rust *Puccinia glumarum* causes regular stripes on the leaf blades. This rust is therefore known as stripe rust.

Stripe of barley (*Hordeum vulgare*) caused by *Helminthosporium graminium.* Long dark brown, linear stripe of leaves; affected leaves turn brown, twist and dry up; no grain formed.

Stripe rust of barely (*Hordeum vulgare*) and wheat (*Triticum aestivum*) Caused by *Puccinia glumarum (P. striformis).* Small elliptical pustules on leaves, which run in rows forming stripes which later turn black; sometimes pustules appear on leaves sheaths and glumes. For control, grow resistant varieties.

Stump rot of tea (*Camellia sinensis*) Caused by *Irpex destruens.* Stump of bushes which have been pruned low and from which lateral branches arise rot forming a cup shaped centre. Apply coal tar to pruned cuts.

Stunting The symptoms can be seen of the mosaic diseases on plants having stunted and dwarfing in size.

Stylospores The spores bearing hyphae are sometimes enclosed in roundish or cup shaped or flask-shaped receptacles known as pycnia or pycnidia and the spores within them are known pycnispores, pycnidiospores or stylospores.

Sugarcane downy mildew Also invades maize, caused by *Perosclerospora sacchari.*

Sulphur dust To control powdery mildew, sulphur dust application is recommended and has shown good result. But the cost of the treatment is prohibitive.

Sulphur fungicides These are very effective fungicides and are used in a number of forms. Any finely divided sulphur may be used as a dust. It should be fine enough to pass through 300 mesh sieve. Ground sublimed and precipitated sulphurs are on the market for this purpose. Lime sulphur is another fungicide and well serves the purpose of an insecticide also.

The important fungicides are sulphur powder, wettable sulphur, lime-sulphur and organo-sulphur. Sulphur powder is a very cheap fungicide, but its efficiency depends upon the size of particles which should be 250 to 300 mroh. Wettable sulphur is used on crops which suffer from lime-sulphur scorching.

Sulphur-lime solution preparation Water (50 gallons) is first boiled in an iron, tin or zinc, but not in copper vessel and then quick lime (50 lbs) and sulphur (100 lbs) are added. The mixture is stirred and boiled for one hour. The liquid is strained and its density taken with a hydrometer; enough water is added to bring the specific gravity to 1.005 or 1.010. Proprietary produce *e.g.* Sulfinette, Limsul, etc. are available in powder from which on dilution are ready for use as lime-sulphur spray. Lime-sulphur spray should not be used on tender foliage as it is likely to scorch it. A number of organo-sulphur compounds *e.g.* thiram, dithiocarbonates, etc. are extremely safe to use on most sensitive foliage. Tiliisan, Ziram are some of the proprietary products based an which organo-sulphur compounds are also used.

Survival The fungi adopt different devices for their survival.

Sweet lime Family *Rutaceae, Citrus limmetoedis.*

Swine fever *See* Animal Science.

Swine pox Pigs of two months of age may be infected with vaccinia either naturally or artificially and show pock lesions on the eyelids, snot, outside the thigh and under surface of the abdomen. The virus of vaccinia may produce an abortive type of pox or grade upto a severe type. The elimination of lice may possible would do a great deal to control this disease.

Symptoms of disease The internal or external reactions or alterations as a result of a disease attack.

Synchytrium In some yeasts, development of mycelium is limited, and in general like Synchytrium the mycelium is completely absent.

Synchytrium collapsum Minute galls caused by this fungus on the leaves of *Clerodendron infortunatum.*

Synchytrium endobioticum Wart of potato. In Phymycetes, in the simplest form of this sub-class a zoospore is the potential gamete, for in *Synchytrium endobioticum,* it has been found that zoospores sometimes swim towards other zoospores and conjugate in pairs, and thus assume the role of gametes.

Synkyron Same as zygote.

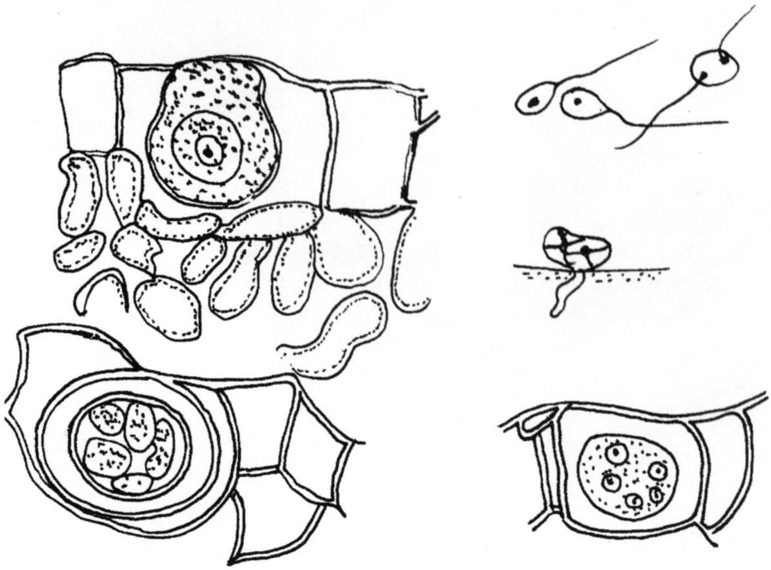

Summmer Sporangia and Resting Sporangia of Synchytrium Showing Fusion of Zoospores (Gametes) and Germination Zygote

Systemic disease *Ustilago holleri,* which causes the covered smut of oats, attacks the host in the seedling stage and keeps on growing with the host without causing any apparent injury. At the time of ear formation, however, the fungus manifests itself when the ovaries are replaced by the sori which contains the spores of the fungus. Such diseases are known as systemic disease.

Systemic mycelium When the mycelium is systemic, it induces the formation of galls, witches' brooms and other malformations in the host in which it perenates.

T

Taphrinaceae The Ascomycetes fungi belonging to this family; of which Taphrina is the most representative genus, are true endoparasites, strictly confined to definite hosts.

Taphrina deformans *See* leaf curl of peach and apricot.

Tar leaf spot of job's tear millet (*Coix lachrymajobi*) Caused by *Phyllachora coicis*. Numerous shiny black blisters appear chiefly on under surface of leaf.

Tar sand *See* oil sand.

Tar spots *See* Phacidiales.

Tea blight (Grey) The causal organism is *Pestalotia thesae;* is the commonest blight of this crop.

Tellitia caries Bunt of wheat, spores rough.

Telletia foetida On wheat, spores are smooth.

Telletia horrida Bunt of rice.

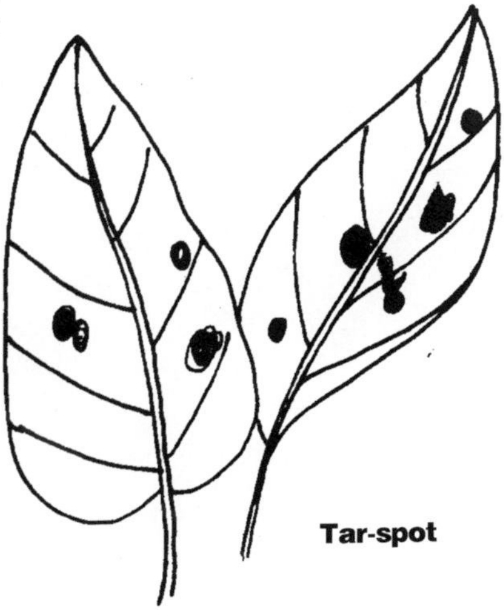

Tar-spot

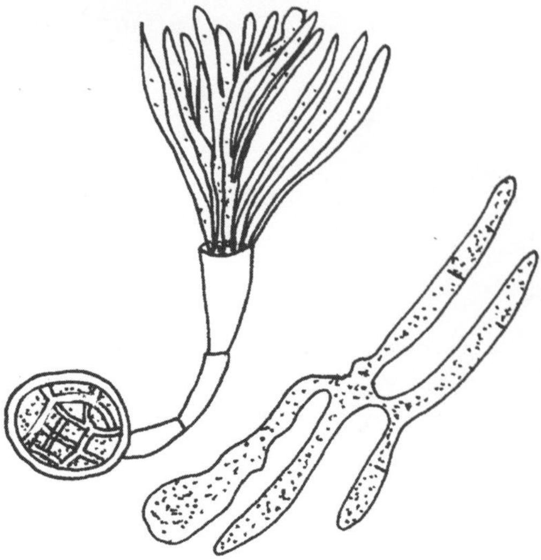

Germination of Tellitia Caries

Temporary parasite A parasite which comes in contact with the host periodically *viz.* mosquito.

Tetanus This is an infectious disease of animals and man, and is characterised by spasmodic tetany and hyperaesthesia. The disease is prevalent all over the world. The causative agent is *Clostridium tetani.* It is a rod shaped anaerobe with rounded ends. It forms a terminal spore which is the twice the width of microorganism and gives an appearance similar to drumstick. The spores are highly resistant and withstand desiccation indefinitely and 5 per cent phenol from 15 hours. It produces a highly potent toxin. It occurs in goats, cattle, pigs, dogs and cats and rarely in poultry. Sometimes the disease develops after a history of wound, surgical interference, shearing, docking or even an injection. Proper hygiene and cleanliness at castration and other surgical procedures should be observed. Also *See Animal* Science.

Tetracoccus Bacteria found as tetrad of four cells.

Thermal death point Somewhat above the maximum temperate the growth of fungi is a temperature exposure to which it is a relatively short period of time. This is the death point of the fungus. The thermal death point may be very high; in *Byssochlamys fulvus*, it is 90°C.

Thiosan A fungicidal dust used as a seed dresser.

Tikka disease of groundnut *(Arachis hypogea* L) used by *Mycosphaerella personata* and the other by *Cercospora arachidiocola* the former is more widespread. The perfect state of *C. persona* is *Mycospherella berkeleyii* and that of *C. arachidicola* is *Mycosphaerella arachidicola.*

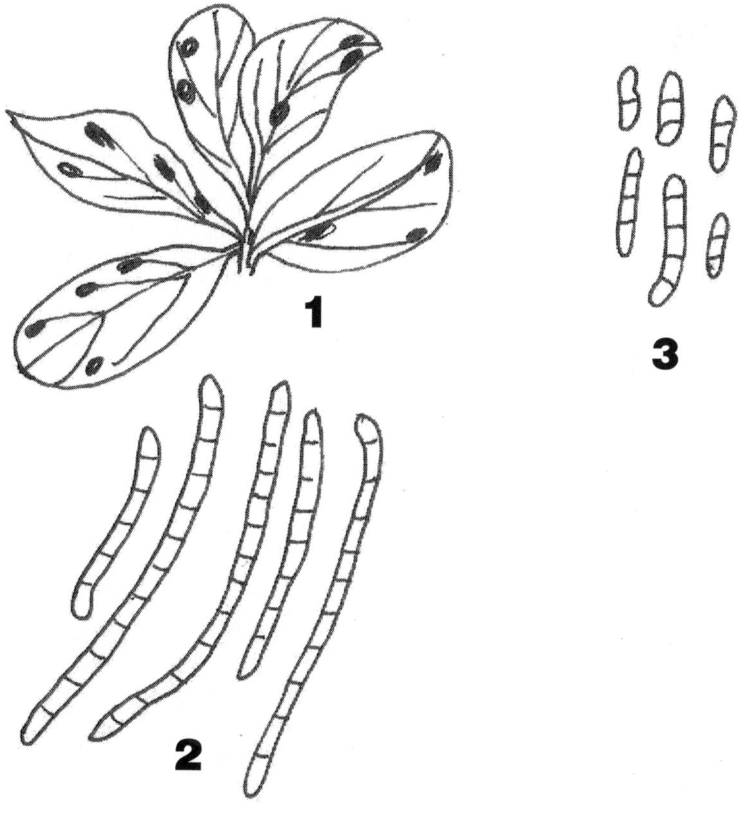

Tikka Disease of Groundnut

(1) Symptoms due to cercospora personata;
(2) Conidia of cercospora arachidicola;
(3) Conidia of *C. personata*

Tikka disease of groundnut (*Arachis hypogea* L.) Caused by *Mycospherella personata.* Necrotic circular spots appear on both sides of leaves; spots on upper surface and dark brown with yellow halo. Spray with 1 per cent Bordeaux mixture; good cultivation and crop rotation; grow resistant varieties.

Tirak disease of cotton (*Gossypium* spp.) It is physiological disease, causes early yellowing, drooping and shedding of leaves, balls crack, do not open properly, seeds remain immature; lint quality greatly reduced. Can be controlled by proper manuring, irrigation and adjusting time of sowing.

Tobacco blank shank The plants are attacked by the organism *Phytophtora nicotianae,* attacks the foot first of the host and brings about sudden death.

Tobacco leaf-curl The leaves get badly curled by this virus and thus interferes with all the physiological functions of the plant.

Tolyposporium cherenbergii Long smut on sorghum.

Tolyposporium penicillariae Smut of bajra *(Pennisetum typhoides).* Also *See* smut of pearl millet *(T. selegalensis).*

Tophrina maculans *See* leaf blotch of turmeric.

Torula Same as sprout cells.

Toxic substances The fungi after invading the plants secrete some toxic substances and kills them.

Trathola flassorbitilis The parasite *Mermis* sp. play an important role in the natural control of the pest.

Transduction Bacteriophages or simply phages are viruses like organisms which invade the bacterial cells multiply therein in ultimately kill or dissolve the host cells–a phenomenon known as lysis.

Transformation In this method of sexual union, the DNA is extracted from dead bacterial cells in laboratory and is then added to a culture of recipients (female cells), the DNA moves from cell to cell in the culture medium and transformation is affected.

Trichoconis padwickii *See* stack bum disease of paddy.

Trichothecium roseum A fungus causing fruit rot apples.

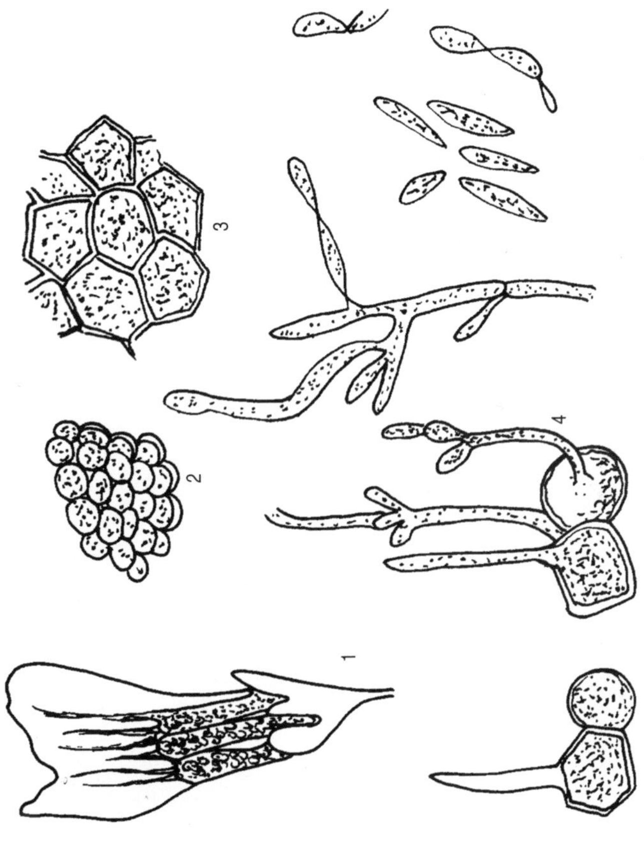

Tolyposporium cherenbergii

1: Ruptured sori; 2: Apore balls; 3: Part of some in section; 4: Germination of spores

Tristeza of citrus Caused by virus. Trees show gradual decline and therefore characterised by profuse flowering and heavy setting of fruits, suppression of new growth, bomzing and yellowing of leaves, defoliation and die back of twigs. Use certified bud wood and suitable stock scion combination. Sweet orange scion seem to be incompatible, with a number of stock, remarkably, with the sour orange, on which they grow for two to three months but die within a year. It is said that sweet orange produces some substances toxic to the sour orange. The 'tristeza disease' is used for this condition.

Tuberculosis *See* Animal Science.

Turmeric (*Curcuma longa*) Suffers badly from the disease of leaf spot caused by *Taphrina maculans*.

U

Udbatta disease of paddy (*Oryza sativa*) Caused by *Balansia oryzae.* Infected ears look like straight spikes, as they come out of leaf sheath, covered with white mycelium which is later transformed into black, hard, dotted sclerotium-like body. It also attacks the Italian millet.

Ufra disease of paddy (*Oryza sativa*) Caused by *Ditylenchus angustus.* Seedling show chlorosis of leaves which later wither and die; initernodes become dark; ears remain covered in leaf sheath and enlarge into spindle-shaped swellings; when severely infected, ear becomes dark brown in colour and is twisted spirally. For control, burn all infected parts and straw.

Ulcerative enteritis A bacterial disease of animals. *See* Animal Science.

Uncinula It is one of the orders of Euascomycetes. *U. necator* is the causal organism of the powdery mildew disease of grapes.

Uncinula necator *See* powdery mildew of grape.

Uredinales These are rusts which are obligate parasites capable of existing on living green plants, principally conifers and flowering plants.

Uredo paspaliscrobiculati *See* rust of kodo millet.

Uredospores Germination of the aeciospores leads to the formation of a diadaryotic stroma within the host from the base of which grow short, two celled upright branches. The upper cell becomes the urediospore and the lower the pedicel. The sori in which these uredospores are formed may not have a distinct or definite shape and are known as uredia. When the air is changed with moisture, and is still, repeated production of uredospores takes place over a period of days. These spores are yellow to orange-red, globose or oval binucleate with their wall made up two distinct layers.

Urocystis brassiceae This organism is as root-galls forming smut.

Urocystis cepulae *See* smut of onion.

Urocystis tritici *See* flag smut of wheat.

Uromyces anthylidis *See* rust of fenugreek.

Uromyces appendiculatus It is an autocious rust, but the pycnidial and aecial stages are not very commonly formed. *See* also rust of cowpeas.

Uromyces decoratus *See* rust of sann-hemp.

Uromyces cicerisarietini *See* under rust of gram.

Uromyces fabae *See* rust of peasand lentil.

Uromyces mucunae *See* rust of velvet bean.

Uromyces setariaeitalicae *See* rust of Italian millet.

Uromyces striatus Rust of alfalfa,

Uromyces trifolii *See* rust of clover.

Ustilago This genus has many diseases caused by various spp. *U. tritici* (loose smut), *U. nuda* (loose smut), *U. hordei* (covered smut), *U. kolleri* (covered smut), *Ustilago scitaminea* (whip smut), *U. avenae* (loose smut); are some of the destructive diseases caused by this genus.

Ustilago scitaminea *See* smut of sugarcane.

Ustilaginales These are parasitic fungi, some species of which are facultative saprophytes. They are restricted to the flowering plants.

Ustilaginoidea virens *See* false smut of paddy.

Ustilago coicis* and *uslachrymijobi *See* smut of job's tear millet.

Ustilago cremeri *See* smut of Italian millet.

Ustilago crussgalli *See* shoot smut of barnyard millet.

Ustilago kolleri *See* covered smut of oats.

Ustilago nuda *See* loose smut of barley.

Ustilago panicifrumentacei* and *U. paradoxa *See* smut of barnyard millet.

Ustulina zonata *See* charcoal rot of tea and jack fruit.

V

Valasaceae A family of sphaerales, of which the asexual state is more prominent state on the living host and perithecia are formed only on dead tissue.

Verum A wilt-resistant variety of cotton.

Verrucosis Name given to sour orange scab-disease.

Vesicle (1) In Pythium, cleavage may not in some species, take place within the sporangium, instead a vesicle is formed at the apex of the sporangium, into which the protoplasm moves; there cleavage takes place and zoospore are formed. (2) During the development of conidia in *Aspergillus niger* the vegetative cell from which a conidiophore is formed–a foot cell. The tip of the conidiophore becomes swollen to form a club-shaped or rounded structure called vesicle.

Virion An infectious virus particle or unit of virus is called virion. The extracellular inert phase of a virus. A virion consists of a protein coat surrounding one or more strands of DNA or RNA. Virions may be polyhedral or helical and vary greatly in size.

Virology It is the study of viruses.

Virulence The relative pathogenecity of an organism. Virulence depends on the invasiveness and toxicity of the pathogens. It may vary between strains of the same organisms or variety of same plant.

Virus A very minute infectious agent that causes a variety of diseases in plants and animals, such as small pox, common cold and various mosaic diseases of plants. Viruses can only reproduce in living cells and outside they can exist as inactive particles surrounded by a protein coat. *See* Mosaic of Cardamom. *See* Leaf curl of Tobacco. *See* Mosaic of Tobacco. *See* Chirke of Cardamom. All the unicellular plants and animals, bacteria and viruses. are known as microorganisms or simply microbes. Modern taxonomists have placed bacteria viruses in group Monera.

Virus diseases In addition to fungi and bacteria, a third group of infectious agents that cause diseases in plants are the viruses.

Virus disease of cotton *See* stenosis disease of cotton.

Volutin The volutin are the granules formed when ribosomes in bacteria, the seat of protein synthesis, the protoplasm contains several types of granules, one being volutin.

W

Warts The outer membrane when sculptured variedly forms warts. *See* endospores also.

Wart of potato (*Solanum tuberosum*) Caused by *Synchytrium endobioticum*. Very small to large blackish protuberances may be observed on under ground stem, stolons and tubers. Grow resistant varieties.

Wettable sulphur *See* sulphur fungicides. By adding certain substances such as glue flour, calcium caseinate, skim milk, a sulphur paste can be made which readily becomes suspended in water. This spray is known as wettable sulphur.

Wet rot or damping off of cotton (*Gossypium* spp.) Causal organism is the fungus *Phytophthora parasitica* and *Pythium* sp. Leaf and collar rot are very promiment.

White blister or white rust of *Brassica* spp. Caused by *Cystopus candidus*. Isolated white shiny pustules developed on leaves and stems; when fully developed, pustules have a powedery consistency; hypertrophy of young stems and flowering parts forming enlarged with distorted organs.

White rust Common disease of cruciferous plants caused by *Cystopus candidus. Albugo candida*. Derives its name formation of white mealy patches on the surface of the infected host.

White rust of *Brassica* spp. *See* white blisters of *Brassica* spp. Also *See* Albugo.

White rust of mustard (*Brassica* spp.) Caused by fungus *Albugo candida.* All parts of plant except roots attacked; white or creamy yellow pustules, variable size and shapes, formed mainly on lower surface of leaf; pustules coalesce and form patches; sometimes stems and flowers are hypertrophied into various types of swellings. Have clean cultivation. Remove and burn affected plant parts.

Spots due to White Rust on Mustard Leaves

Wilts These are primarily the diseases of roots, as the causal organism attack them in the seedlings stage.

Wilt of banana Caused by *Fusarium oxysporium* var. *cubense.* Progressive browning and falling of leaves; black streak in underground stem. For control, eradicate affected plants; apply crude oil to form; select disease free suckers, practise crop rotation; treat soil with nematicide.

Wilt of black pepper (*Piper nigrum*) Caused by *Pythium* spp. The vine wilts at or below ground level. Good drainage essential.

Wilt of chillies (*Capsicum annum*) Caused by *Fusarium annum.* Cortical decay of collar and roots. Avoid heavy soils; give irrigation in furrows.

Wilt of coconut Caused by virus. General wilting of the leaves, flaccidity and ripping of leaflets and necrosis of leaf tips; yellowing of old leaves; the crown becomes smaller due to progressive reduction in the number of size of the leaves. Roots are affected and they dry from the tip backwards.

Wilt of cotton (*Gossypium* spp.) Caused by *Fusarium vasinfectum.* Leaves become yellow and droop; plants stunted, wilts and dies;

vascular tissues become brown. Grow resistant varieties for control of diseae.

Wilt of lathyrus Caused by *Fusarium arthoceros* var. *lathyri.* Chlorosis of leaves and discolouration of vascular tissue of roots; plants wither and dry.

Wilt of linseed (*Linum usitatisssimum*) Caused by *Fusarium lini.* Plant die young due to attack of fungus; older plants become yellowish, remain stunted and finally die. Grow resistant varieties; treat seed with Agrosan GN before sowing @ 2 gms/kg.

Wilt of maize Caused by *Fusarium moniliformis.*

Wilt of pigeon pea (*Cajanus cajan*) Causal organism is *Fusarium indica.* Leaves become yellowish, droop down, affected branches wither and dry up; roots become dark brown, are devoid of rootless and show a black streak below the bark. Grow resistant varieties.

Wilt of pyrethrum (*Chrysanthemum cinarariaefolum*) Caused by *Phytophthora cambivora* and *Rhizoctonia solani.* Progressive drying of the leaves and shoots.

Wilt of sann-hemp (*Crotoloria juncea*) *Fusarium udum* var. *crotolarae.* Leaves turn prematurely yellow and wither; vascular tissues of roots and stem are discoloured black; entire plant dries up. For control, grow resistant varieties; practise crop rotation and crop sanitation.

Wilt of sugarcane (*Saccharum officinarium*) Caused by *Cephalosporum sacchao.* Leaves dry up; stem becomes light and hollow; pith develops purple or red streaks. Grow resistant varieties; use healthy seed.

Wilt of sesamum (*Sesamum orientale*) Caused by *Fusarium vasinfectum.* Same as in cotton.

Winter streak *See* mosaic of tomato.

Witches' brooms Woody plants sometimes produce closely grouped clusters through the formation of irregular tufts by their slender branches. The branches from which such tufts develop may become swollen. They are turned upwards and form hypertrophied leaves. These disfigurations in many cases are produced by parasitic fungi whose mycelium infests the internal tissues of the plant. These structures are the witches' brooms.

Witches broom of peach and apricot (*Prunus persica* and *Prunus armeniaca*) Caused by manganese deficiency. Main vein and mid-rib of leaf remain green, while interveinal and peripheral areas are chlorotic. Prune off witches broom; spray tree with manganese sulphate solution (1 lb. in 50 gallons of water).

X

Xanthomonas alfalfae *See* bacterial disease of alfalfa or lucern.

Xanthomonas compestris *See* black rot of crucifers.

Xanthomonas camopsidis *See* bacterial spot of cluster beans.

Xanthomonas malvacearum *See* black arm of cotton.

Xanthomonas oryzae *See* bacterial blight of paddy.

Xanthomonas phaseoli indicus *See* bacterial blight of Phaseolus.

Xanthomonas phaseoli sojens *See* bacterial disease of soyabean.

Xanthomonas sesbaniae *See* bacterial disease of Sesbania.

Xanthomonas stizolobiicola *See* bacterial leaf spot of velvet bean.

Xanthomonas vasculorum *See* Gummosis of sugarcane.

Xanthomonas vesicatoria *See* bacterial leaf spot of chillies.

Xanthomonas vignica *See* bacterial blight of cowpea.

Y

Yeast The fungi belonging to the family *Saccharomycetes*, are known as yeasts, which are used in making bread, because of their ability to lighten the loaf by filling it with bubbles of carbon dioxide and in the production of alcohol by the fermentation of sugar.

Yellow *See* infectious chlorosis.

Yellow mosaic of *Phaseolus* spp. Caused by virus. Affected plants show yellow and green areas on leaves; severely affected leaves turn completely yellow.

Yellow leaf spot of sugarcane (*Saccharum officinarum*) Caused by *Carospora kopkei*. Leaves show yellowish spots.

Yellow mosaic of lablab bean (*Dolichos lablab*) Light and dark green patches on leaves in advanced stages, the leaves become yellow.

Yellow rust of barley and wheat *See* stripe rust of barley and wheat.

Yellow rust of wheat Same as stripe rust of wheat–*Puccinia glumarum*.

Yellow or stripe rust of wheat and barley Caused by *Puccinia striiformis*.

Yellow vein mosaic of lady's finger (*Abelmoschos esculantus*) Caused by virus. Yellowing of leaves. Grow resistant varieties.

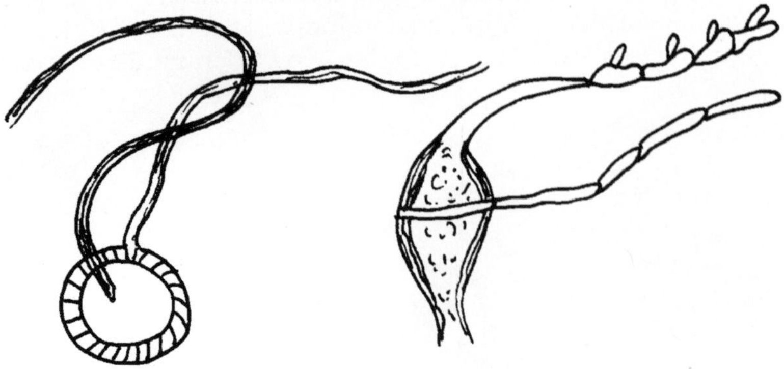

Yellow Rust of Wheat
a: Germinating uredospore; b: Germinating teleutospore

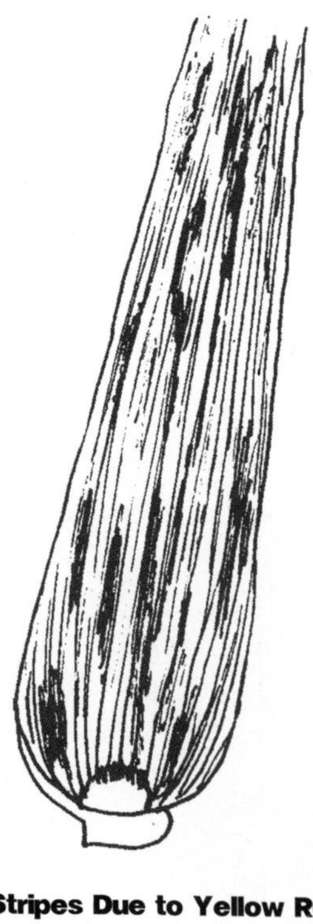

Stripes Due to Yellow Rust

Yellowing rot of wheat This disease is characterised by curling of stalks and inflorescence and transformation of the spikelets into a bright-yellow, slimy mass due to a bacterium, *Phytomonas tritici.*

Yellowing disease of sugarcane (*Saccharum officinarum*) Caused by physiological condition. Leaves turn yellow from tip downwards and dry up leaving mid-rib green; later whole crown droops and dies. Apply nitrogenous fertilizers alone or with phosphate.

Ysigenous *See* schizogenous or *See* schizonesis.

Z

Zygosporangium The fusion nuclei of the zygospore in *Mucor* after having undergone meiotic division before the start of germination and the promycelium contains nuclei of two strains *i.e.* plus and minus. This promycelium bears at its tip an asexual sporangium called zygosporangium.

Zoospore In Algae, the cells of the filament except the holdfast are capable of producing zoospore which are liberated into a delicate mucilagenous vesicle through a small lateral aperture on one side of the cell.